Cultural Heritage Science

More information about this series at http://www.springer.com/series/13104

Kaori Fukunaga

THz Technology Applied to Cultural Heritage in Practice

Springer

Kaori Fukunaga
National Institute of Information
and Communications Technology
Tokyo, Japan

ISSN 2366-6226 ISSN 2366-6234 (electronic)
Cultural Heritage Science
ISBN 978-4-431-55883-5 ISBN 978-4-431-55885-9 (eBook)
DOI 10.1007/978-4-431-55885-9

Library of Congress Control Number: 2016932883

Printed on acid-free paper

This Springer imprint is published by Springer Nature
The registered company is Springer Japan KK

Preface

This book introduces currently available terahertz (THz) technology that can be useful for heritage science by showing case studies of museum objects. Heritage scientists can overview prospects and limitations of THz spectroscopy and THz time-domain imaging by comparing well-established classic methods. It is hoped that some may think of adding the THz region in the frequency range that they use. Conservators and historians can examine whether this new technique is already useful or is not quite ready for them. This book can be informative for THz researchers who look for applications that are actually used by people in different fields.

The book contains six chapters. There are many books and review papers that describe instruments, starting from semiconductor laser devices; therefore, this book omits information about how to generate and detect THz waves. Chapter 1 describes the current status of THz technology with some examples in industrial applications. Chapter 2 introduces turnkey-type THz systems available in the market because we users have a little intention to build systems by ourselves. Chapter 3 introduces the possible use of THz spectroscopy for material analyses in heritage science, by showing spectra of pigments, binders, and mixtures. When it comes to imaging, almost all THz-related papers, including review papers for imaging, do not explain how to obtain a false colour image from a set of pulses. It is essential to understand the process of making the image to avoid misinterpretation. Thus, Chap. 4 explains the process of THz pulsed time domain imaging (TDI) in detail. THz pulsed TDI results are compared with other common methods such as X-ray radiography in Chap. 5. This chapter suggests something unique in the use of THz waves. Finally, various case studies, including examination of masterpieces, for example, a tempera panel painting by Giotto and panel screens by Ogata Korin and Kano Eitoku, are introduced in Chap. 6. This chapter characterizes the book, the title of which ends with “in practice”.

The author hopes this book will encourage research on signal analyses of THz imaging and collaborations between THz experts and conservation scientists to develop and improve THz technology for cultural heritage.

Tokyo, Japan Kaori Fukunaga

Acknowledgments

The application of THz technology for art conservation science began about 10 years ago, and the work introduced in this book has been supported by numbers of people.

First of all, the author thanks Yuichi Ogawa of Kyoto University for his help in THz spectroscopy by using his FT-THz system, as well as artists and conservators for making samples by traditional art materials. They include Takasuke Nishiyama, Satoshi Hamaya, Yoshimi Ueno, and especially Miho Bokuda, who currently works in Florence as an independent conservator. Their collaboration without evidence of the potential of THz technology for art allowed the author to go forward.

The author expresses sincere thanks to the conservators Stefano Scarpelli, Rita Alzeni, and Laura Lucioli; historians and curators Cristina Accidini, Antonio Natali, Angelo Tartufelli, Francesca de Luca and their colleagues in Polo Museale Fiorentino as well as Maria Pia Zaccheddu and Don Ottavio Failli for providing the opportunities to observe masterpieces in Italy and for fruitful discussions on experimental results.

The author expresses special appreciation for sincere collaborations in Japan to conservators and artists Misao Ohno, Akiko Okimoto, Haruhiko Suzuki, Kikuko Iwai, Yasunori Oka and his colleagues of Oka Bokkodo; historians and curators Toru Tateishi and Shigemasa Udagawa of Agency for Cultural Affairs Japan, Akira Tomita and Kie Hanzawa of Tokyo Station Gallery; and the temples and shrines Kiyomizu-dera in Kyoto, Tanzan shrine in Nara, and Aizendo in Osaka.

The author is most grateful to researchers working in heritage science for their support and discussions: Marcello Picollo and his colleagues of Istituto di Fisica Applicata "Nello Carrara" – Consiglio Nazionale delle Ricerche, Susanna Bracci and her colleagues of Istituto per la Conservazione e la Valorizzazione dei Beni Culturali – Consiglio Nazionale delle Ricerche, Matija Styrlic and his colleagues in University College London, Michel Menu and his colleagues in Centre de Recherche et de Restauration des Musées de France, Marco Leona and his colleagues in The Metropolitan Museum of Art New York, Bernhard Blümich and his colleagues in Aachen University, Yohsei Kohdzuma and his colleagues in the Nara

National Research Institute for Cultural Properties, Nobuyuki Kamba and his colleagues in the Tokyo National Museum, Yoshihisa Fujii and his colleagues in Kyoto University, as well as former Ph.D. students including Kazuyuki Yamamoto, Tanja Trafela, Julien Laboune, and Corinna Ludvica Koch Dandolo. Additional thanks go to researchers in nondestructive testing, including Takahide Sakagami and his students in Kobe University.

The applications to heritage science cannot be realized without the help of THz experts. The author is grateful to researchers, including J. Bianca Jackson and her colleagues in Reading University, Emmanuel Abraham and his colleagues in Université du Bordeaux, Gian Piero Gallerano and his colleagues in Ente per le Nuove Tecnologie, l'Energia e l'Ambiente, Frascati, Irl N. Dulling III and his colleagues in Advanced Photonics Inc., Tomofumi Ikari and his colleagues in Spectra Design, Phillip F. Taday and his colleagues in TeraView Ltd., Naoki Oda and his colleagues in NEC Corp., and Masakazu Ogasawara and his colleagues in Pioneer Corp.

My passion for art grew naturally thanks to my parents and brother, who regularly took me to museums and galleries from infancy; and my sincere thanks go to Christian Laurent and Hisako Niko for their continuous encouragement to write this book.

The main research fund for this work was given by the National Institute of Information and Communications Technology (NICT) by the decision of a former president, Hideo Miyahara. Some of the field measurements were also funded by the Agency for Cultural Affairs Japan. Finally, the author would like to express the greatest appreciation to Iwao Hosako and all colleagues at NICT.

Contents

Chapter 1
Introduction

Electromagnetic waves are widely used in nondestructive testing (NDT). The operating frequency region determines the role of the waves, because the responses of materials depend on the frequency, as shown in Fig. 1.1. The penetration depth and the spatial resolution are also determined by a combination of the frequency and the object's physical properties.

1.1 Electromagnetic Waves Used in Heritage Science

Every frequency band has its own advantages, and heritage scientists have been using all available frequency regions. For example, ground-penetrating radar in the microwave region has been used to find heritage objects or ruins that are buried underground. In the optical frequency range above the infrared region, electromagnetic waves have been used for material analysis and for imaging based on the material information. Fourier transform infrared spectroscopy (FTIR) is commonly used as a chemical analysis tool in various industries, and has been widely used to identify materials by micro-sampling. Noninvasive material analysis is also possible using fibre optic reflectance spectroscopy (FORS), which has been applied to investigating various masterpieces. The most practical use of the IR region could be for IR and NIR (near-infrared) cameras. These cameras can detect carbon black with high sensitivity, thus enabling drawings hidden under paints and faded texts written using carbon black to be visualised. Certain organic dyes and varnishes can be examined using ultraviolet (UV) light when they exhibit specific responses, although the UV light may affect the object by causing photochemical reactions. The X-ray is also a type of electromagnetic wave, although it is an ionising radiation. In addition to traditional X-ray radiography, X-ray fluorescence (XRF) and X-ray diffraction (XRD) methods are commonly used for element analysis applications, such as mineral pigment analysis [1–9]. These electromagnetic wave-based scientific analysis methods for heritage science have been studied since the

K. Fukunaga, *THz Technology Applied to Cultural Heritage in Practice*, Cultural Heritage Science, DOI 10.1007/978-4-431-55885-9_1

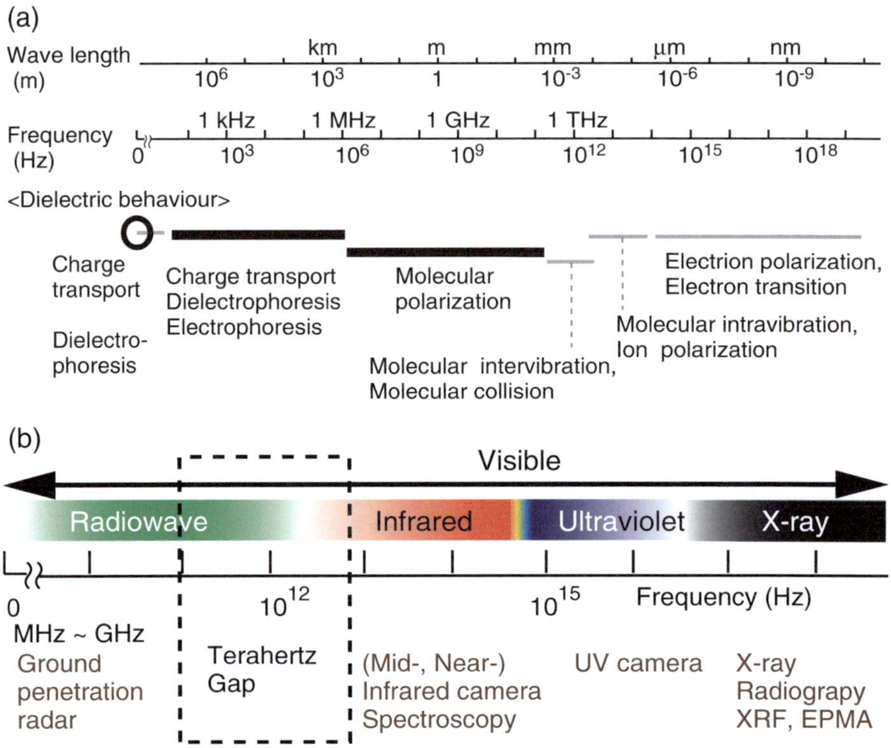

Fig. 1.1 Use of electromagnetic waves in materials investigation. (**a**) Dielectric response of materials, (**b**) spectroscopy and imaging techniques

1960s, and are well established in optical frequency range. The waves in this terahertz (THz) gap in the band are also called submillimetre waves and far-infrared rays. THz spectroscopy and imaging are among the emerging technologies in heritage science; international projects have included its applications [10, 11], and some recent review papers on heritage science cover THz applications [12, 13].

1.2 Terahertz Waves

The frequency range of THz waves typically refers to the 0.1–10 THz range (1 THz = 300 μm, and 33.3 cm^{-1} = 4.2 meV, 48 K), which lies at the border between the photonic and electronic ranges. Previously, this frequency range had only been used for specialized academic works, such as astrophysics and semiconductor physics [14, 15]. Limited numbers of researchers have been using THz waves with in-house or custom-made THz spectroscopy systems since the 1960s. THz spectroscopy can be used to investigate materials science phenomena such as molecular behaviour, including polarisation, inter-vibration and intra-vibration, as shown in Fig. 1.1a. Because the energy of the THz waves corresponds to weak bonds, such as the hydrogen bond, THz spectroscopy is expected to become a useful analysis tool for organic-inorganic composites, including biomaterials. In addition to classical THz spectroscopy using a Michelson interferometer, i.e., the far-infrared version of the common FTIR method, THz time domain spectroscopy (TDS) technology was developed in the 1990s. A THz-TDS system uses narrow pulses with pulse widths of ps order. By applying a fast Fourier transform (FFT) to the transmission pulse in the time domain, spectra in the frequency domain are obtained. While THz spectroscopy is used in practical applications in the pharmaceutical industry [16–19], theoretical analysis of THz spectroscopy has not yet been performed adequately because of the complicated phenomena involved, such as simultaneous relaxation and resonance in dielectric materials [20, 21]. THz spectroscopy of art materials is discussed in greater detail in Chap. 3.

1.3 Terahertz Applications to NDT in General

While THz spectroscopy is not yet commonly used in most industries, the THz-TDS technique has been accepted as a new nondestructive test (NDT) method that filled the gap shown in Fig. 1.1b. The imaging results that were published in the mid-1990s encouraged research and development towards industrial applications, especially in security applications for concealed dangerous object detection [22, 23]. Thanks to recent developments in laser technology and commercial systems, which are introduced in Chap. 2, THz technology has been applied to the inspection of products that require high production quality standards, such as materials for the space industry. A number of review papers and books have been published to introduce proposed industrial applications of the technology [24–31].

Imaging techniques using THz pulses in the time domain can show the internal structures of opaque objects based on the time-of-flight technique. Figure 1.2 shows a schematic diagram of the penetration depth of electromagnetic waves for a typical painting model. The THz waves can penetrate more deeply than infrared light, and can obtain structural information from dielectric materials that are relatively transparent to X-rays. These are major advantages of the use of THz waves in NDT applications, which can be applied not only to the imaging of paintings, but also to the imaging of various multi-layered industrial products, e.g., laminated roofing materials.

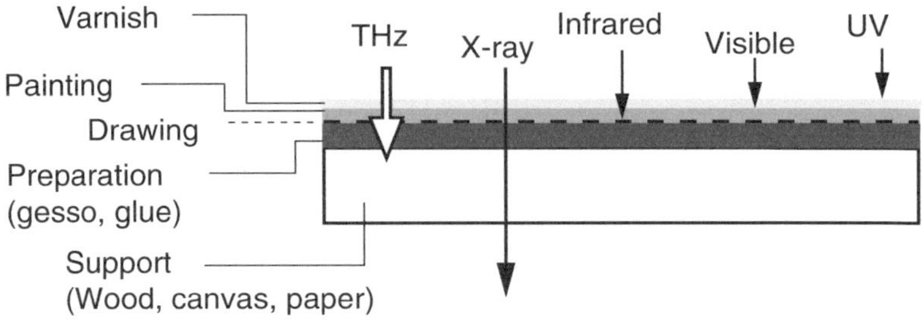

Fig. 1.2 Typical painting model and penetration depths of electromagnetic waves

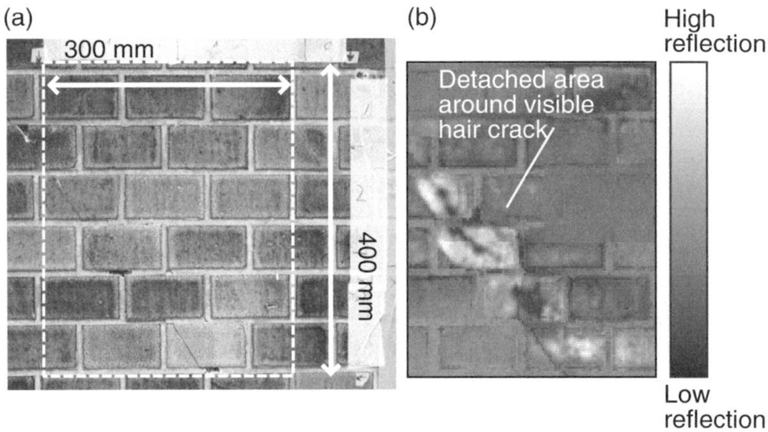

Fig. 1.3 Example of defect detection in tiles on a wall in a concrete building. (**a**) Observation area on wall, (**b**) THz reflection image

The technique has several names, including terahertz pulsed imaging (TPI), THz time domain imaging (THz TDI), and THz pulse-echo imaging. Because "pulse" and "time-domain" are essential aspects of this technique, this book uses "THz pulsed TDI". The technique is described in detail in Chap. 4, and a comparison of THz pulsed TDI when applied to cultural heritage objects with the properties of existing NDT methods introduced in Chap. 5.

Before the application of these techniques to cultural heritage is discussed, some industrial applications are described here to illustrate the potential of THz imaging techniques. Figure 1.3 shows the detection of air gaps between tiles and mortar layers in a building after a major earthquake. A highly reflective area was detected along a visible hairline crack, likely caused by an air gap between the tiles and the mortar on the concrete wall. Details of this example are discussed along with the results that were obtained by microwave and millimetre wave continuous-wave (CW) imaging systems in Chap. 5.

One of the most promising examples of the use of this technology is detection of corrosion under paint (CUP), which is an important topic in industrial NDT, and in

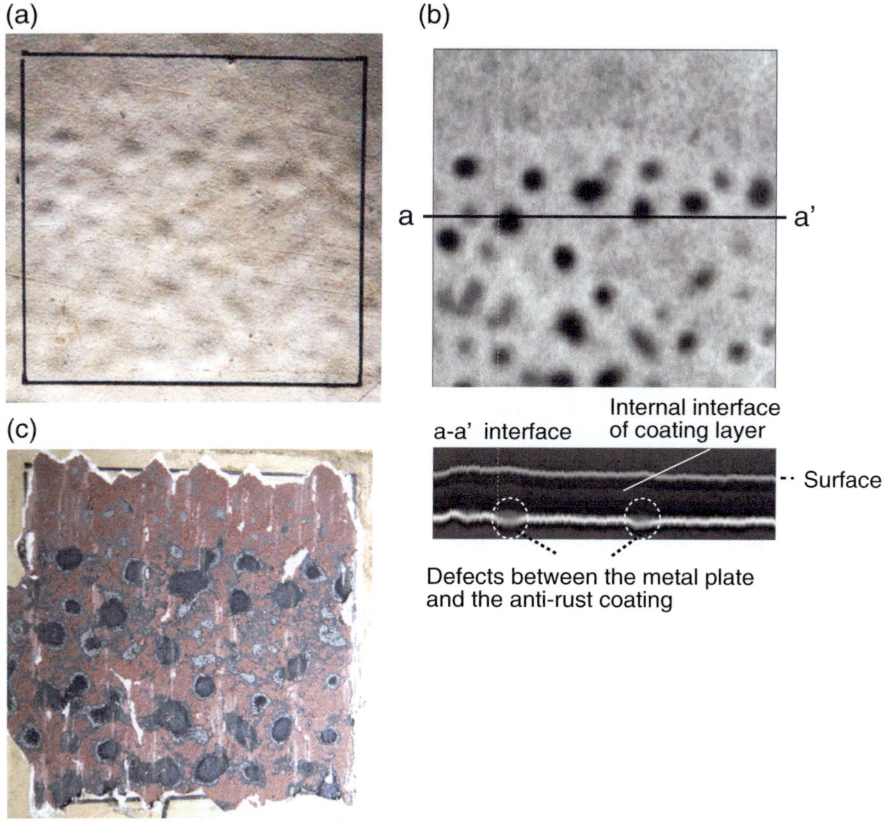

Fig. 1.4 Observation of corrosion under paint in an oil storage tank. (**a**) Observation area, (**b**) THz reflection image and cross section along the line a-a′. The layer structure of the anti-rust coating is clearly observed to include the presence of defects between the metal plate and the coating, (**c**) confirmation by destructive test by removal of the coating, performed after the THz pulsed TDI examination

archaeological applications [32–34]. The reflection level of an oxidised metal is much lower than that of a new metal, which means that corroded areas can be detected easily. According to recent work, the progress of corrosion can be monitored using THz pulsed TDI and the method is thus expected to be used for diagnosis applications. Figure 1.4 shows an example of the observation of CUP in an oil storage tank. There are many small lumps on the floor of the tank, as shown in Fig. 1.4a. The THz area image and cross-section, which are shown in Fig. 1.4b, revealed that corrosion existed under most of these lumps, and showed the presence of a dielectric material that has a different refractive index to that of anti-corrosion paint.

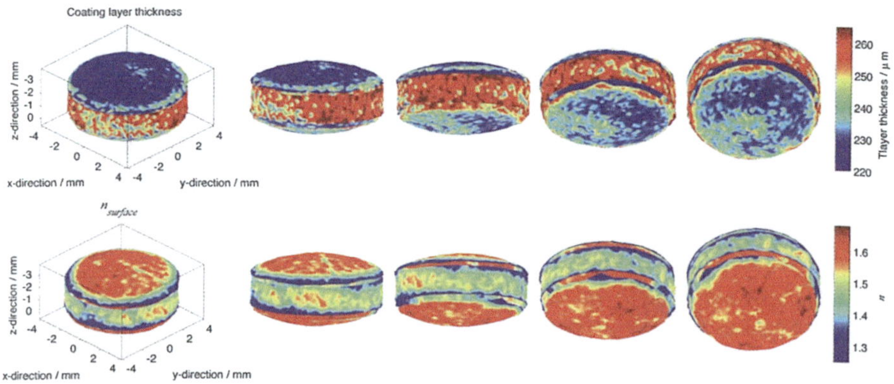

Fig. 1.5 Example of THz pulsed TDI of pharmaceutical tablet [31, p. 474, Fig. 18.18]

As shown in these examples, the surface conditions under paint can be observed quite easily. However, the client may desire automatic diagnoses based on the image, and that requires the accumulation of huge numbers of case studies.

In the pharmaceutical industry, material mapping and structural observations of medicines have been performed based on spectroscopic imaging. Figure 1.5 shows an example of THz pulsed TDI of a medical tablet. The colour map indicates the substance distribution and the three-dimensional image is used for quality control of the tablet shape. If each art material has a specific feature in the THz frequency region, then similar mappings could be obtained for art materials.

The structural observations of industrial products made using polymeric materials also indicate the potential of THz pulsed TDI for application to modern heritage objects that commonly contain synthetic materials [35–38].

1.4 Attempted THz Applications in Heritage Science and Their Progress

Early attempts to apply THz technology to artworks were aimed at spectroscopic imaging, including the detection of hidden paintings and materials identification. The results can be displayed as false colour images, similar to those used in infrared imaging techniques [39–42]. However, other useful classical methods are available for surface analysis applications, and the trial measurements were not very successful. THz pulsed TDI was first applied to a tempera masterpiece at a museum in 2008 to observe the internal structures of the preparation layers, and the clear cross-section images showed that it was painted by following the medieval technique. The results are discussed in detail in the case study in Sect. 6.3 [43]. Many research groups around the world have actively used these techniques in recent years. Experimental results for various types of objects, including panel paintings, wall

Table 1.1 Examples of artworks studied using THz technologies

Objects	Types	Examples [T] Transmission [R] Reflection
Art materials	Pigments, dyes Binders, Supports (wood, canvas, etc.)	Spectral analysis by FTIR in 1970s [T]
		Spectral database by FTIR and TDS [T] [R]
		Defects detections in wood support by TDI [R]
		Tree ring analysis by TDI [R]
Paintings	Mural (fresco)	Spectral analysis of paints by TDS [R]
	Panel (tempera)	Structure observation of layers by TDI [R]
	Canvas (oil)	Defects detection by TDI [R]
Vases, other 3D objects	Wood, Clay, etc.	Structure observation by CW [T] and TDI [R] NB. Tomography technique is required
Documents	Papyrus, Clay, Parchment	Spectroscopic imaging by TDS [T] [R]
		Layer extraction by TDI [R]
Decoratives	Panel screen	Spectroscopig imaging by TDS [R]
	Interior walls	Structure observation by TDI [R]
Mummies	Wrapped	Structure observation by TDI [R]
	Unwrapped	Observation of bones by CW [T] and TDI [R]

FTIR Fourier transform infrared spectroscopy
TDS time domain spectroscopy
TDI THz pulse time domain imaging
CW continuous wave

paintings, and archaeological vases, have proved that the THz frequency band has several advantages for nondestructive internal observation of artworks [44–47]. Table 1.1 gives a list of examples, including the model experiments.

In this type of work, public interest may arise following the discovery of hidden paintings or drawings under top paintings, for which the discussion is limited to the layers near the surface. In such a case, the image resolution would become critical. While the analysis of painting layers is important in art history studies, observation of the preparation layer condition is also useful for conservation planning. Among the many benefits of using THz technology, noninvasive cross-sections that can be taken at positions of interest and area images taken at the depths of interests are among the greatest. The practical examples that are described in the following chapters will hopefully encourage the use of THz technology in heritage science.

References

1. S. Keck, Appl. Opt. **8**, 41 (1969)
2. C. Gaffney, J. Gater, *Revealing the Buried Past: Geophysics for Archaeologists* (Tempus, Stroud, 2003)
3. D. Pinna, M. Galeotti, A. Mazzeo, *Scientific Examination for the Investigation of Paintings. A Handbook for Conservator-Restorers* (Centro Di, Firenze, 2009)

4. M. Leona, R.V. Duyne, Chemistry and Materials Research at the Interface Between Science and Art. Report of a NSF and A.W. (Mellon Foundation, 2009)
5. J.R. Wiseman, F. El-Baz, *Remote Sensing in Archaeology* (Springer, Heidelberg, 2007)
6. M. Bacci, R. Bellucci, C. Cucci, C. Frosinini, M. Picollo, S. Porcinai, B. Radicati, Fiber optics reflectance spectroscopy in the entire VIS-IR range: A powerful tool for the non-invasive characterization of paintings. MRS Proceedings, 852 (2004)
7. F. Casadio, L. Toniolo, J. Cult. Herit. **2**, 71 (2001)
8. R.J.H. Clark, Appl. Phys. A. **89**, 833 (2007)
9. A. Anitha, A. Brasoveanu, M. Duarte, S. Hughes, I. Daubechies, J. Dik, K. Janssens, M. Alfeld, Signal Process. **93**, 592 (2013)
10. INSIDDE project. http://insidde-fp7.eu
11. Charisma Project. http://www.charismaproject.eu
12. M. Alfeld, J.A. Broekaert, Spectrochim. Acta. B. **88**, 211 (2013)
13. A. Pelagotti, A. Del Mastio, A. De Rosa, A. Pica, IEEE Signal Process. Mag. **25**, 27 (2008)
14. P.H. Siegel, IEEE Trans. Anntenas Propag. **55**, 2957 (2007)
15. K.L. Nguyen, T. Friscic, G.M. Day, L.F. Gladden, W. Jones, Nat. Mater. **6**, 206 (2007)
16. J. Pearce, D.M. Mittleman, Phys. Med. Biol. **47**, 3823 (2002)
17. P.F. Taday, I.V. Bradley, D.D. Arnone, M. Pepper, J. Pharm. Sci. **92**, 831 (2003)
18. P. Taday, Philos. Trans. A. **362**, 351 (2004)
19. C.J. Strachan, P.F. Taday, D.A. Newnham, K.C. Gordon, J.A. Zeitler, M. Pepper, T. Rades, J. Pharm. Sci. **94**, 837 (2005)
20. Y. Ohki, M. Okada, N. Fuse, K. Iwai, M. Mizuno, K. Fukunaga, Appl. Phys. Express **1**, 122401 (2008)
21. M. Scheller, S. Wietzke, C. Jansen, M. Koch, J. Phys. D Appl. Phys. **42**, 065415 (2009)
22. B.B. Hu, M.C. Nuss, Opt. Lett. **20**, 1716 (1995)
23. D.M. Mittleman, R.H. Jacobsen, M.C. Nuss, IEEE Sel. Top. Quants Electron. **2**, 679 (1996)
24. D.M. Mittleman, M. Gupta, R. Neelamani, R.G. Baraniuk, J.V. Rudd, M. Koch, Appl. Phys. B. **68**, 1085 (1999)
25. D.M. Mittleman, *Sensing with terahertz radiation* (Springer, Berlin, 2003)
26. D.L. Woolard, E.R. Brown, M. Pepper, M. Kemp, Proc. IEEE **93**, 1722 (2005)
27. W.L. Chan, J. Deibel, D.M. Mittleman, Rep. Prog. Phys. **70**, 1325 (2007)
28. R.A. Cheville, Terahertz time-domain spectroscopy with photoconductive antennas, in *Terahertz Spectroscopy. Principles and Applications*, ed. by S.L. Dexheimer (CRC Press, Taylor and Francis Group, Boca Raston, 2008), pp. 1–39
29. M. Tonouchi, IEEE Terahertz Sci. Tech. **2**, 90 (2009)
30. C. Jansen, S. Wietzke, O. Peters, M. Scheller, N. Vieweg, M. Salhi, N. Krumbholz, C. Jördens, T. Hochrein, M. Koch, Appl. Opt. **49**, 48 (2010)
31. J.A. Zeitlerm, Y. Chun Shen, "Industrial applications of teraherts imaging", chapter 18 of Kai-Erik Peiponen, Makoto Kuwata-Gonokami, in *Terahertz Spectroscopy and Imaging*, ed. by J. Axel Zeitler (Springer-Verlag, Berlin, 2013), pp. 451–489
32. N. Fuse, T. Fukuchi, T. Takahashi, M. Mizuno, K. Fukunaga, IEEE Trans. THz Sci. Tech. **2**, 242 (2012)
33. A. Moriguchi, T. Tanaka, T. Sakagami, M. Hangyo, Proc. 13th Intern. Symp. on Nondestructive Characterization of Materials, **44** (2013)
34. J.B. Jackson, J. Labaune, G.A. Mourou, L. D'Alessandro, A. Whyte, M. Menu, Proc. 36th Intern. Conf. IRMMW-THz and Terahertz Waves, (2011)
35. S. Wietzke, C. Jördens, N. Krumbholz, B. Baudrit, M. Bastian, M. Koch, J. Eur. Opt. Soc. Rapid Publ. **2**, 07013 (2007)
36. J.-W. Park, K.-H. Im, D.K. Hsu, C.-P. Chiou, D.J. Barnard, Adv. Mater. Res. **123–125**, 839 (2010)
37. C.-C. Chen, D.-J. Lee, T. Pollock, J.F. Whitaker, Opt. Express **18**, 3477 (2010)
38. G. Pastorelli, T. Trafela, P.F. Taday, A. Portieri, D. Lowe, K. Fukunaga, M. Strlič, Analy. Bioanal. Chem. **403**, 1405 (2012)

39. W. Köhler, M. Panzner, U. Klotzbach, E. Beyer, S. S. Winnerl, M. Helm, F. Rutz, C. Jördens, M. Koch, H. Leitner, Proceedings of the 9th ECNDT, **181** (2006)
40. K. Fukunaga, Y. Ogawa, S. Hayashi, I. Hosako, IEICE Electron. Express **4**, 258 (2007)
41. J.B. Jackson, M. Mourou, J.F. Whitaker, I.N. Duling, S.L. Williamson, M. Menu, G.A. Mourou, Opt. Commun. **281**, 527 (2008)
42. E. Abraham, A. Younus, J.C. Delagnes, P. Mounaix, Appl. Phys. A. **100**, 585 (2010)
43. K. Fukunaga, Non-destructive THz pulsed TDI of Giotto masterpiece. News Conserv. **10**, 2 (2009)
44. K. Fukunaga, M. Picollo, Appl. Phys. A. **100**, 591 (2010)
45. J.B. Jackson, J. Bowen, G. Walker, J. Labaune, G. Mourou, M. Menu, K. Fukunaga, IEEE Trans. Terahertz Sci. Tech. **1**, 220 (2011)
46. K. Fukunaga, M. Picollo, Characterisation of works of art, in K. E. Peiponen et al. eds. *Terahertz Spectroscopy and Imaging*, Springer Series in Optical Sciences, vol. 171, (Springer, Berlin, 2013), pp. 521–538
47. E. Abraham, K. Fukunaga, Stud. Conserv. **60**, 343 (2015)

Chapter 2
THz Instruments

Various instruments that use THz technologies are commercially available. The operating frequencies span from approximately 0.1 to 20 THz. In this chapter, THz instruments are introduced in terms of two main application categories: spectroscopy and imaging.

2.1 FTIR-Based Spectroscopy Systems

In the past decade, the main interest in THz technology in the scientific community is focused on time domain systems. However, it should be noted that conventional Fourier transform THz spectroscopic (FT-THz) systems that use Michelson interferometers (Fig. 2.1a) can also be successfully applied to the analysis of materials [1, 2]. FT-THz technology, including the software for analysis, is well established with the progress in mid-infrared spectroscopy techniques. However, actual material identification is not yet possible because of the lack of a spectral library and the lack of the theoretical background in materials science in the THz region.

Figure 2.1b shows a compact FT-THz system (VIR-F, JASCO, Tokyo, Japan) with a ceramic or mercury lamp source and a pyroelectric triglycine sulfate (TGS) detector: this device makes it possible to collect transmittance or reflectance spectra in the 1.0–20 THz range (40–680 cm^{-1}) with a resolution of 30 GHz at room temperature. The system uses a silicone beam splitter that allows broadband spectroscopy to be performed [3, 4]. Because the absorption peaks of water/moisture in the THz region are not negligible, the entire optical path of the device is kept in a vacuum. The specimen is placed in a "sample compartment" with a linear dimension of 10 mm; this compartment is the only part of the device that is open to the air. Figure 2.1c shows a widely distributed product (Vertex series, Bruker Optics, Ettlingen, Germany) that offers a far-infrared option. This system covers the frequency range from the THz region to the mid-infrared (5–50,000 cm^{-1}). Figure 2.1d shows one of the more classical instruments of this type (DA8,

K. Fukunaga, *THz Technology Applied to Cultural Heritage in Practice*, Cultural Heritage Science, DOI 10.1007/978-4-431-55885-9_2

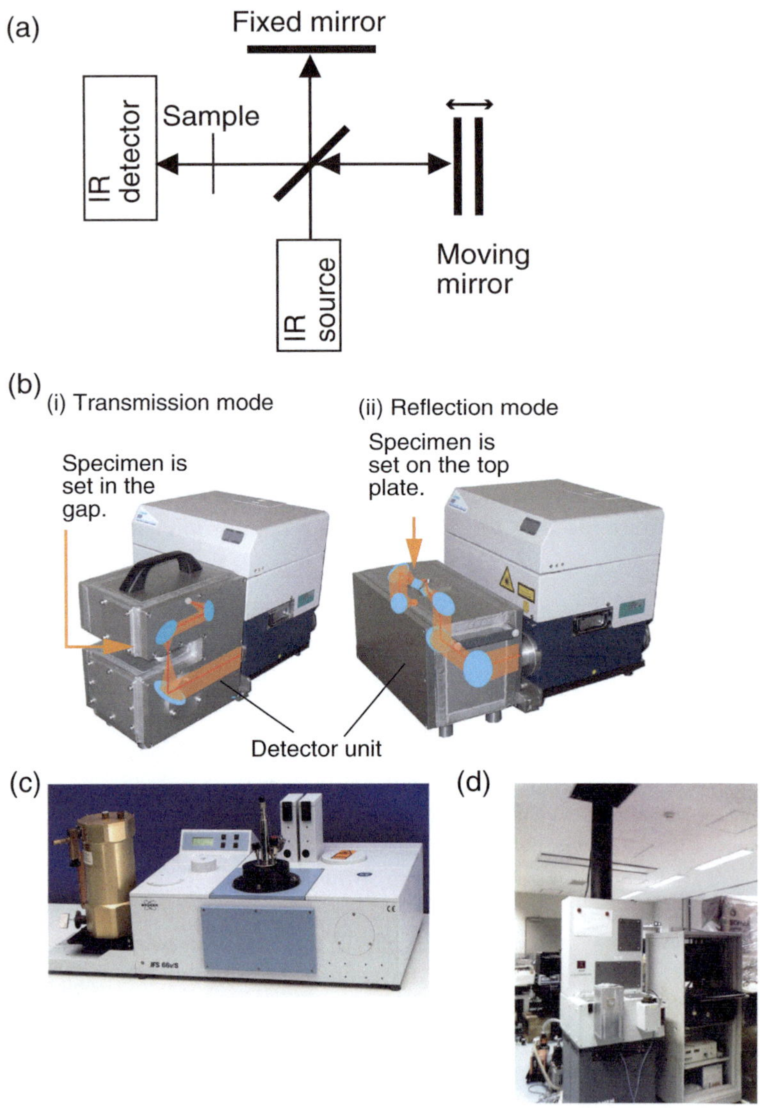

Fig. 2.1 FT-THz systems: principles and examples of commercial systems. (**a**) Schematic of the principle of the FT-THz system, (**b**) JASCO VIR-F, (**c**) Bruker Vertex Series, (**d**) ABB Bomem DA8

Bomem ABB Analytical, Quebec City, Canada), which has high resolution of 0.15 GHz, is suitable for applications in the range from 0.9 THz to the mid-infrared region (4–50,000 cm^{-1}), and is equipped with a broadband beam splitter for the frequency range from 40 to 700 cm^{-1} (1.2–21 THz) at room temperature. This system has been used to observe the absorption spectra of gases, mainly by researchers in astronomy and meteorology, and to analyse nonlinear crystals [5, 6].

Recently, integrated systems operating from the mid-infrared to the far-infrared have been developed, and this type of approach may encourage chemists and materials engineers to use the THz frequency range [7].

2.2 Time Domain Spectroscopy Systems

Since the 1990s, research activities in the THz frequency range have been dominated by the development of THz TDS systems. The most common products of this type use a photoconductor-based antenna as a source. Irradiation of this antenna using a femtosecond optical pulse from a fibre laser generates the THz pulse. Another antenna is used as the detector. All aspects of THz TDS, from devices to entire systems, are explained in detail in various books and review papers [8–11].

Figure 2.2a shows a schematic diagram of the TDS method in transmission mode. This configuration is generally used for spectroscopy and for simple imaging without depth profiling, such as the detection of a metallic object in an envelope, which is used to demonstrate the method. The time gate is swept using an optical delay line. The terahertz waveform is then generated from the collected signals. The reflection mode, which is shown in Fig. 2.2b, is the most common configuration used in THz pulsed TDI for on-site experiments in cultural heritage applications. The emitter and the detector are enclosed in a sensor unit, and the position of the unit is numerically controlled. The sensor unit configuration effectively reduces the time required for system maintenance, such as alignment of the optical path. Figure 2.2c shows the attenuated total reflection (ATR) mode, which is now widely used in common FTIR systems for mid-infrared spectroscopy applications. This mode is suitable for use in laboratory experiments on liquid specimens [12].

Figure 2.3 shows examples of commercially available TDS systems. Figure 2.3a shows one of the oldest products in Japan (Rayfact, Tochigi Nikon, Tochigi, Japan), which allows modifications to be made by the researchers. Figure 2.3b shows a product (TeraPulse 4000, TeraView, Cambridge, UK) that was developed by a company that has collaborated with pharmaceutical companies for many years, and Fig. 2.3c shows an example of a product (Terasys 4000, Rainbow Photonics, Zurich, Switzerland) that was developed by a spin-off company from an academic institution. Various types of THz-TDS systems are available on the market, although there is no standard evaluation method for these systems. Even the power measurement method is not internationally standardised. The Physikalisch-Technische Bundesanstalt (PTB) in Germany has started a TDS system calibration service based on basic metrological research in the optical research field [13, 14]. The National Institute of Information and Communications Technology (NICT) in Japan provides a practical user guide for sample fabrication, noise elimination, and evaluation of uncertainty [15]. The National Physical Laboratory in the UK is conducting inter-comparison tests on THz-TDS systems in collaboration with standards institutions, including the National Institute of Standards and Technology (NIST) in the USA and the NICT. This international activity should contribute to the establishment of system validation protocols for THz TDS systems.

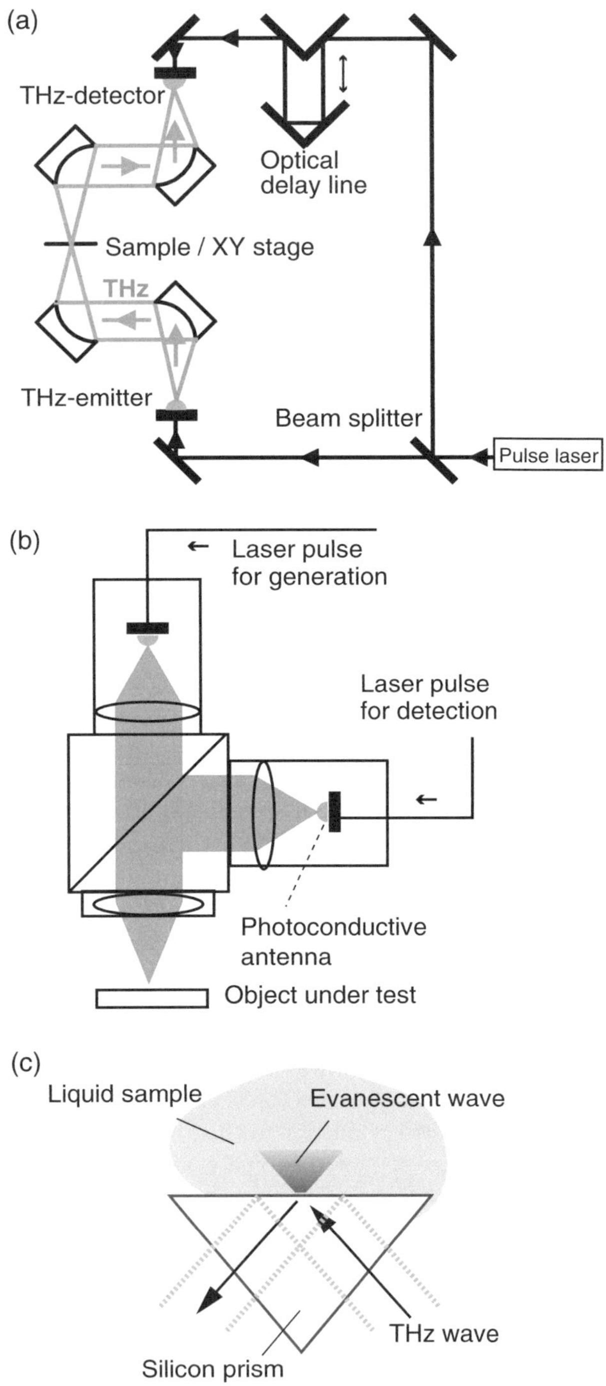

Fig. 2.2 Schematic of principles of THz time-domain spectroscopy and imaging systems. (**a**) Entire system in transmission mode, (**b**) sensor unit for a typical THz time domain imaging system in reflection mode, (**c**) attenuated total reflection (ATR) mode

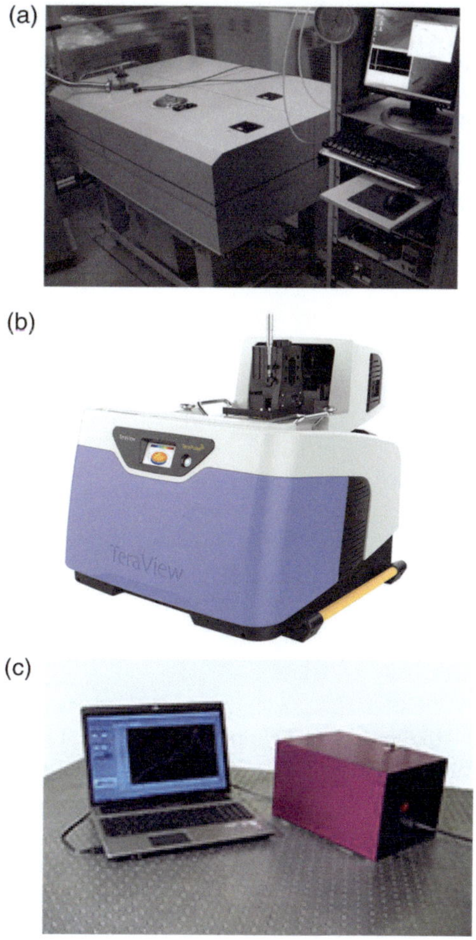

Fig. 2.3 Examples of commercially available THz time domain systems. (**a**) Rayfact, Tochigi Nikon Co., (**b**) TeraPulse 4000, TeraView Ltd., (**c**) Terasys 4000, Rainbow Photonics

It is very important for potential users of these systems to perform as many test measurements as possible, and to compare data obtained using different products.

The THz TDS, which uses a narrow-band pulse (of picosecond order), has the advantage that the refractive index is calculated directly without using the Kramers-Kronig transformation, because the incident and detected pulses include both amplitude and phase information. This does raise the question of who is likely to need the refractive index in the THz frequency region. The clear answer to this question is that the refractive index is needed by the people who develop THz instruments, including telecommunications systems, to select the appropriate materials. For heritage science applications, the spectra that are obtained from the output signals, as shown in Fig. 2.4, are important because they allow us to clarify whether or not this very limited frequency range is useful for the characterization of art materials.

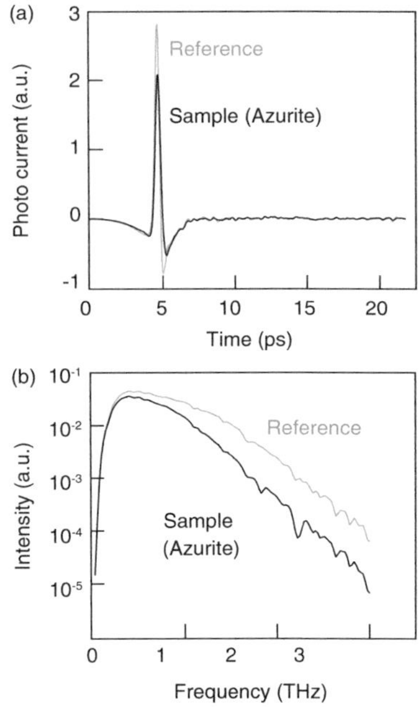

Fig. 2.4 Principle of time domain spectroscopy using an azurite. (**a**) Output signal in time domain, (**b**) signal in frequency domain after a fast Fourier transform (FFT)

In the case of THz pulsed TDI, the THz pulse is used directly in the time domain. While spectroscopic imaging is of course possible, in most cases in practice, the THz pulse echo can provide sufficient information to decide whether a more precise analysis is required or not.

This section briefly introduces a commercially available THz pulsed TDI system along with other in-house imaging systems that are under development. THz pulsed TDI is the main method used in all case studies introduced in this book, and the method is described in detail in Chap. 4, including the prospects of spectroscopic imaging.

Figure 2.5 shows the principle of THz pulsed TDI schematically; theoretical explanations of the method can be found in the references that were introduced earlier [8–11]. When a THz pulse is applied to a double-layered object made from materials A and B, the first reflection is generated at the interface between air and material A, followed by the reflection at the internal interface between materials A and B, and finally the reflection from the surface of material B appears, as shown in Fig. 2.5a. An example of the raw output signal is shown in Fig. 2.5b. The delays correspond to the thickness and physical properties of each material, and this THz pulse sequence indicates the depth profile of the object, thus resulting in “nondestructive cross-section imaging”. Figure 2.5c, d shows an example of a portable THz pulsed TDI system (T-Ray 4000, API Ltd., Ann Arbor, MI, USA) and a photograph of an on-site measurement performed at a conservation studio in Florence, respectively. This system obtains the depth profile at each scanning point, which means that a three-dimensional image can be reconstructed, if necessary [16–18].

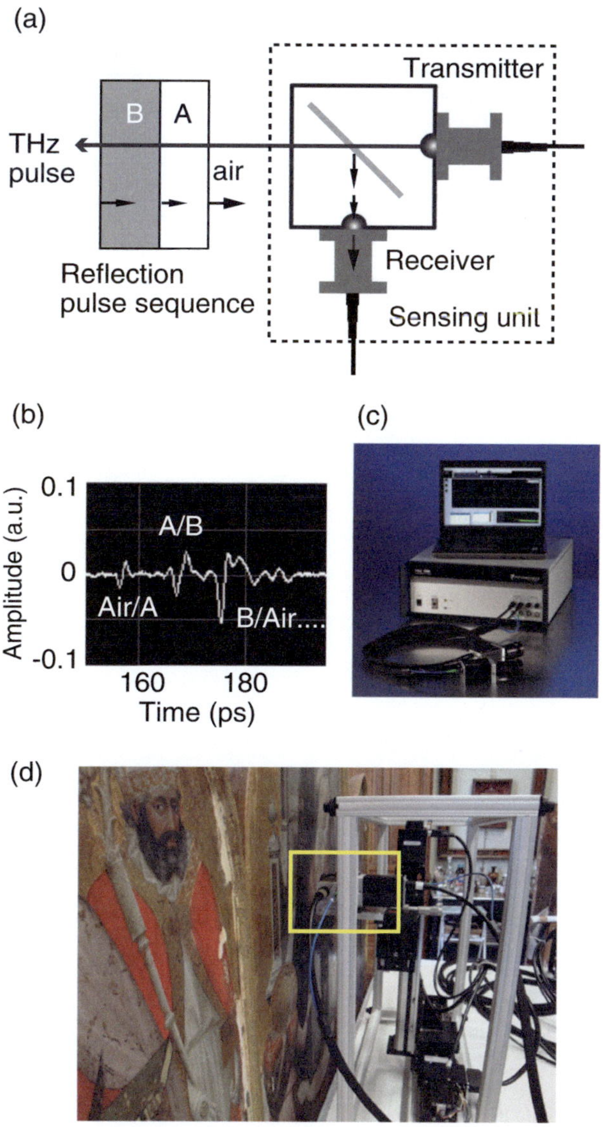

Fig. 2.5 Schematic of TDS imaging principle. (**a**) THz pulse emitted into a double-layered object, (**b**) example of a THz reflection pulse waveform from a double layered object, (**c**) example commercial system: the T-Ray 4000 by API Ltd., Ann Arbor, MI, USA, (**d**) example of on-site measurement setup

2.3 THz CW Imaging and Tomography

In addition to THz pulsed TDI, there are also imaging systems that use continuous waves (CW), including THz tomography systems, and the THz camera that is desirable for use in on-site measurements.

THz CW imaging systems, which operate at approximately 0.1 THz, have major potential for NDT applications. A portable THz imager, in which the THz source is a Gunn diode, was distributed by a German company called SynView [19, 20]. This system has been applied in the observation of mummies, and was also applied to

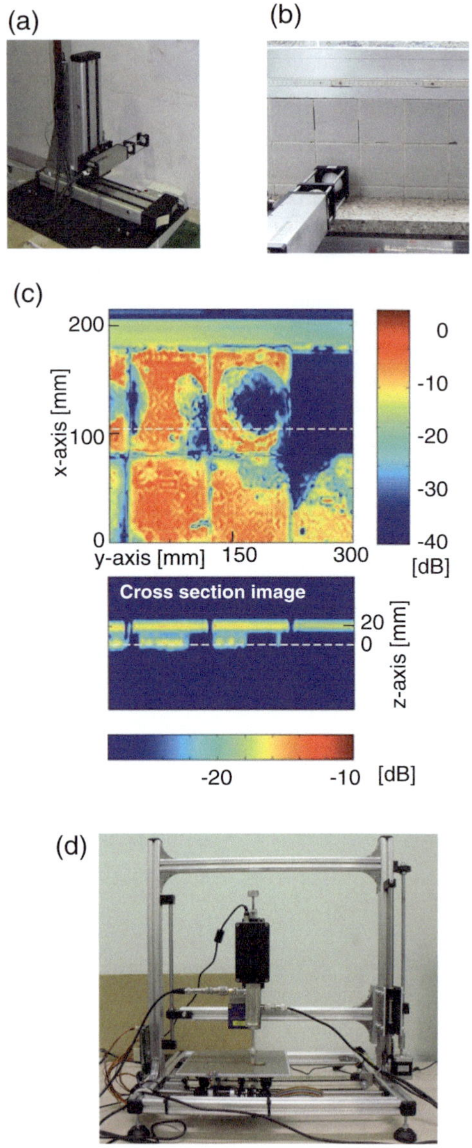

Fig. 2.6 Example of 100 GHz CW imaging system. (**a**) Photograph of system developed for field experiment by the former SynView GmbH, (**b**) observation area, (**c**) reflection image at 100 GHz, (**d**) 97 GHz CW imaging system developed by ENEA Frascati

wall inspection of buildings after the recent major earthquake in Japan, as shown in Fig. 2.6a, b. The nonuniform areas that appeared in the area image and depth profile in Fig. 2.6c clearly indicated air gaps between tiles and mortar. While the company has since ceased its activity in this area, these results confirmed that the frequency range around 0.1 THz is suitable for the analysis of wall paintings with thick mortar layers, such as fresco paintings. ENEA Frascati developed a new THz CW imaging system operating at 97 GHz, which is shown in Fig. 2.6d, using an impact ionization avalanche transit-time (IMPATT) diode, as part of a bilateral collaboration between Italy and Japan called THz-ARTE [21].

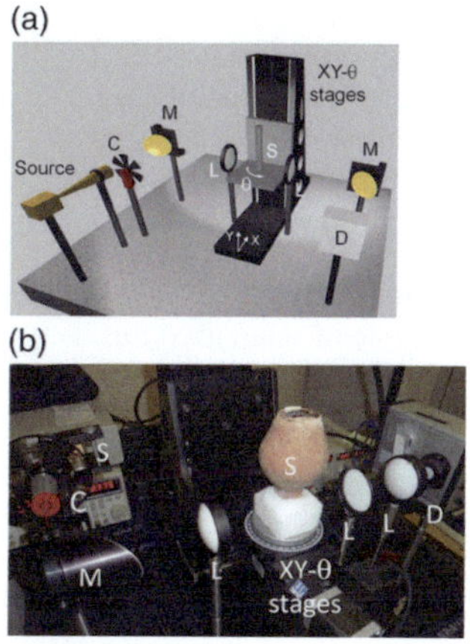

Fig. 2.7 THz computed tomography (CT) system. (**a**) Schematic view of the setup (*C* chopper, *M* parabolic mirror, *L* Teflon lens, *S* sample, *D* pyroelectric detector), (**b**) THz scanner analysing Egyptian pottery in the Museum of Aquitaine (the second parabolic mirror was replaced by a Teflon lens in this case)

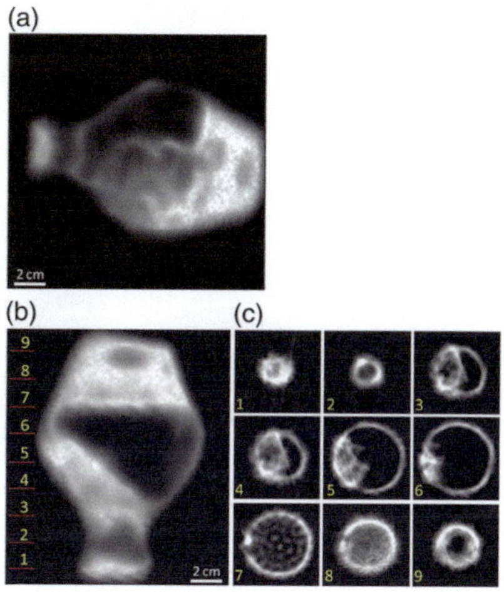

Fig. 2.8 THz CT results. (**a**) Front view, horizontal position, (**b**) front view, upside-down vertical position, (**c**) axial views at position indicated in (**b**)

Bordeaux University has developed a THz CW tomography system (THz scanner) using a Gunn diode source operating at 110 GHz. A schematic diagram of the system and a photograph of an on-site measurement in a museum are shown in Fig. 2.7. The sample is set on a turntable in this configuration, and is then scanned by the THz CW system. The system has successfully looked through sealed Egyptian vases, as shown in Fig. 2.8. The internal object could not be observed by X-ray radiography [22, 23].

2.4 THz Real-Time Imager (THz Camera)

Figure 2.9 introduces a THz camera (IR/V-T0831, NEC, Tokyo, Japan) in which the detector device is based on vanadium oxide micro-bolometer arrays. The THz source, which includes a quantum cascade laser (QCL), is commercially available

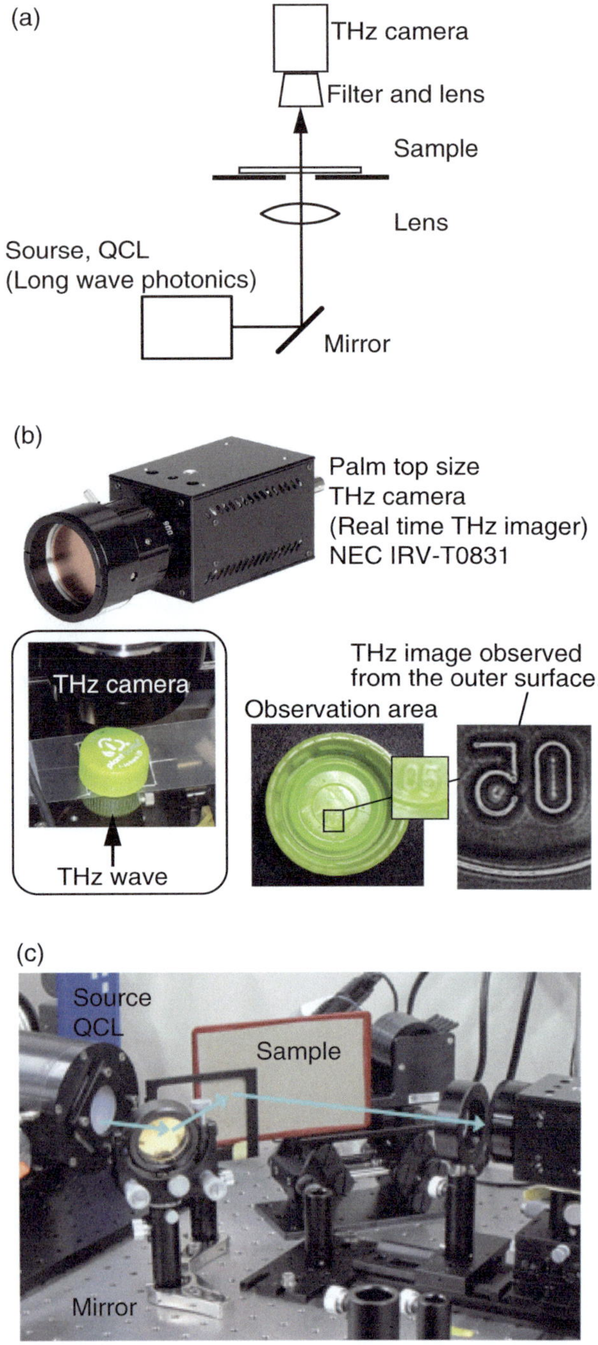

Fig. 2.9 Example of THz camera system. (**a**) Schematic diagram of THz transmission imaging using a THz camera with QCL source, (**b**) example of commercial THz camera developed by NEC, (**c**) example of setup used for reflection imaging by THz camera

from Long Wave Photonics (Mountain View, CA, USA). Figure 2.7a shows a schematic of the configuration for real-time THz imaging in transmission mode. A photograph of the camera and a demonstration of real-time imaging of a number hidden in a plastic cap is shown in Fig. 2.7b. Figure 2.7c shows an example of the reflection mode that is essential for wall painting observation applications. A THz tomography technique using the QCL is also under development, and the development of another interesting application of QCLs, the THz confocal microscope, is in progress. While it is not essential to perform measurements at frame rates of 50 and 60 Hz in the case of museum objects, these rates may be useful when performing rough surveys of large areas, such as entire walls in churches or caves [24–27].

References

1. P.R. Griffiths, J.A. De Haseth, *Fourier Transform Infrared Spectrometry* (John Wiley & Sons, New Jersey, 2007), pp. 112–118
2. M. Quack, F. Merkt (Eds.), *Handbook of High-resolution Spectroscopy* (John Wiley & Sons, New Jersey, 2011)
3. C.C. Homes, G.L. Carr, R.P.S.M. Lobo, J.D. LaVeigne, D.B. Tanner, Appl. Opt. **46**, 7884 (2007)
4. K. Abe, S. Hayashi, N. Doki, C. Otani, K. Kawase, T. Miyazawa, Y. Ogawa, Measurement of hydrated water in D-glucose powder using THz-wave spectroscopy. Bunseki Kagaku **56**, 851–856 (2007)
5. Bruker Optics, VERTEX Series technical note, available from www.bruker.com/optics
6. D.M. Slocum, E.J. Slingerland, R.H. Giles, T.M. Goyette, Atmospheric absorption of terahertz radiation and water vapor continuum effects. J. Quant. Spectrosc. Radiat. Transf. **127**, 49–63 (2013)
7. P. Kühne, C. M. Herzinger, M. Schubert, J. A. Woollam, T. Hofmann, An integrated mid-infrared, far-infrared, and terahertz optical Hall effect instrument. Rev. Sci. Instrum. **85**, 071301 (2014)
8. M. Tonouchi, Galore new applications of terahertz science and technology. Terahertz Sci. Technol. **2**, 90–101 (2009)
9. K.-E. Peiponen, M. Kuwata-Gonokami, J. Axel Zeitler (eds.), *Terahertz Spectroscopy and Imaging* (Springer-Verlag, Berlin, 2013), pp. 451–489
10. D Saeedkia (Eds), *Handbook of Terahertz Technology for Imaging, Sensing and Communications* (Woodhead Publishing, Cambridge, 2013)
11. H-J. Song, T. Nagatsuma, (Eds), *Handbook of Terahertz Technologies; Devices and Applications* (Pan Stanford, 2015)
12. D. Suhandy, M. Yulia, Y. Ogawa, N. Kondo, L-Ascorbic acid prediction in aqueous solution based on FTIR-ATR terahertz spectroscopy. Eng. Agric. Environ. Food **5**, 152–158 (2012)
13. L. Werner, H.-W. Hübers, P. Meindl, R. Müller, H. Richter, A. Steiger, Towards traceable radiometry in the terahertz region. Metrologia **46**, 160–164 (2009)
14. A. Steiger, M. Kehrt, C. Monte, R. Müllwe, Traceable terahertz power measurement from 1 THz to 5 THz. Opt. Express **21**(12), 14466–14473 (2013)
15. M. Mizuno, (Ed.), Practical Users Guide of THz TDS system. in Japanese, available from http://www.nict.go.jp/out-promotion/data-provided/thz_practice_guide.html (2015)
16. D. Zimdars, J.S. White, G. Stuk, A. Chernovsky, G. Fichter, S. Williamson, Large area terahertz imaging and non-destructive evaluation applications. Insight **48**, 537–539 (2006)

17. J. Labaune, J.B. Jackson, K. Fukunaga, J. White, L. dÁlessandro, A. Whyte, M. Menu, G. Mourou, Investigation of Terra Cotta artefacts with terahertz. Appl. Phys. A **105**, 5–9 (2011)
18. I.N. Duling, D. Zimdars, Terahertz imaging: revealing hidden defects. Nat. Photonics **3**, 630–632 (2009)
19. A. Keil, T. Hoyer, J. Peuser, H. Quest, T. Loeffler, *All-Electronic 3D THz Synthetic Reconstruction Imaging System*, Proc. 35th IRMMW-THz (2011)
20. T. Hoyer, T. Loffler, T. Saito, N. Yukihira, A. Deninger, K. Fukunaga, *A Portable All-Electronic THz Scanner for the Inspection of Structural Earthquake Damage in Japanese Buildings*, Proc. 37th IRMMW-THz (2013)
21. A. Doria, E. Giovenale, G. P. Gallerano, M. Picollo, K. Fukunaga, *A Millimeter Wave/Terahertz 3D Scanner for Wall Painting Investigation*, Proc. 38th IRMMW-THz (2014)
22. B. Recur, A. Younus, S. Salort, P. Mounaix, B. Chassagne, P. Desbarats, J.-P. Caumes, E. Abraham, Investigation on reconstruction methods applied to 3D terahertz computed tomography. Opt. Express **19**, 5105–5117 (2011)
23. J.-P. Caumes, A. Younus, S. Salort, B. Chassagne, B. Recur, A. Ziéglé, A. Dautant, E. Abraham, Terahertz tomographic imaging of XVIIIth Dynasty Egyptian sealed pottery. Appl. Opt. **50**, 3604–3608 (2011)
24. K. Fukunaga, N. Sekine, I. Hosako, N. Oda, H. Yoneyama, T. Sudoh, Real-time terahertz imaging for art conservation science. J. Eur. Opt. Soc.: Rapid Publ. **3**, 08027 (2008)
25. N. Oda, T. Ishi, S. Kurashima, T. Sudou, M. Miyoshi, T. Morimoto, T. Yamazaki, T. Tsuboi, T. Sasaki, Palm-size and real-time terahertz imager; and its application to development of terahertz sources, Proc. SPIE Vol. 8716, No. 871603 (2013)
26. A.W.M. Lee, T.-Y. Kao, D. Burghoff, Q. Hu, J.L. Reno, Terahertz tomography using quantum-cascade lasers. Opt. Lett. **37**, 217–219 (2012)
27. R.U. Siciliani de Cumis, J.-H. Xu, L. Masini, R. Degl'Innocenti, P. Pingue, F. Beltram, A. Tredicucci, M.S. Vitiello, P.A. Benedetti, H.E. Beere, D.A. Ritchie, Terahertz confocal microscpy with a quantum cascade laser source. Opt. Express **20**, 21924–21931 (2012)

Chapter 3
THz Spectroscopy

The frequency band below the mid-infrared region was studied actively in the 1960s and 1970s during the development of FTIR spectroscopy. While researchers explored the use of this "far-infrared" region for spectroscopy at that time, only the mid-infrared region has become commonly used in FTIR methods.

3.1 Far-Infrared Spectroscopy

Mid-infrared spectroscopy has progressed rapidly since 1960s, and has become a commonly used tool for chemists. Standard measurement protocols, spectral libraries of standard chemicals, and data formats for common databases have been established by users and manufacturers [1]. Various spectral libraries exist for the mid-infrared region for art materials, including an online database composed by world-leading institutions in heritage science [2, 3]. The far-infrared region, however, has not been commonly used, as described in Chap. 2.

A rich variety of research activities took place during the early stages of FTIR development in the 1960s and 1970s [4–7]. For example, Fig. 3.1 shows the spectrum of the mercury sulphide-based pigment cinnabar, which shows specific sharp and broad peaks that were generated by phonon absorption [5, 6]. This spectrum will be compared with a more recently obtained spectrum later in this chapter. However, frequencies in the range below approximately 12 THz (400 cm^{-1}) have not been used in industrial applications, while the mid-infrared range has been used widely by chemists. Most FTIR instruments cover the range from 12 to 120 THz (400–4,000 cm^{-1}). This may be because the important features of organic materials appear in the mid-infrared region, and these features are sufficient for the polymer industries that progressed so remarkably in the twentieth century. In inorganic materials, however, X-ray related measurements such as XRF and XRD are more useful for the analysis of elements and crystalline structures. In addition to the presence of the other established measurement methods described above, the

K. Fukunaga, *THz Technology Applied to Cultural Heritage in Practice*, Cultural Heritage Science, DOI 10.1007/978-4-431-55885-9_3

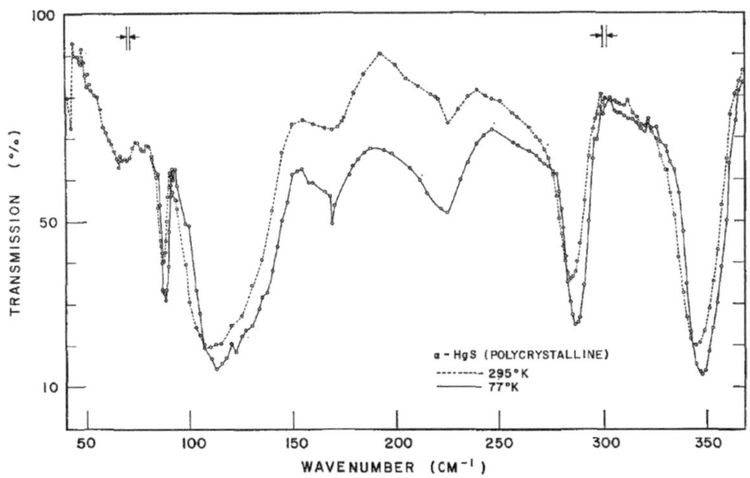

Fig. 3.1 Transmission spectra of polycrystalline trigonal mercury sulfide showing lattice absorption bands at 295 and 77 K (Reprinted by permission from [6]

lack of theoretical investigation related to spectral assignment was another reason for the abandonment of research in THz spectroscopy. THz spectroscopy is not actually new at all; it was simply not required in industrial applications at that time. Even if the spectra in the THz region are only used by academics and are not particularly useful in industrial applications at present, it is still worth looking at spectroscopy in the THz region from a reference perspective.

3.2 Spectroscopy in Transmission and Reflection by FTIR and TDS

When either a CW or pulsed electromagnetic wave is applied to an object, some part of the incident wave is reflected at the surface while the rest penetrates into the material (Fig. 3.2).

The amplitude and phase of the reflected wave are determined by the reflective index of the material. At a metal surface, in theory, 100 % reflection with a phase opposite to that of the incident wave should be obtained. In the dielectric material case, the transmitted wave is used to characterise the material. If no reflection occurs at the surface and the material is homogeneous, the transmitted wave is directly related to the absorption that corresponds to the refractive index. However, in most cases, the transmitted wave is obtained after surface reflection, internal absorption by the substance, and internal scattering caused by the inhomogeneity of the material, as shown in Fig. 3.2. Therefore, the reflective index calculated from

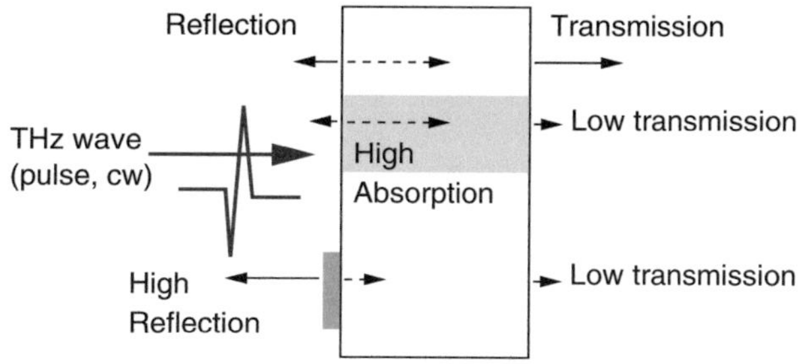

Fig. 3.2 THz pulse reflection and propagation. The transmitted signal is affected not only by absorption, but also by reflection at the surface. Therefore, low transmission does not directly indicate high material absorption

the output signal means the "apparent" value, which is influenced by the sample preparation process and other uncertain or variable conditions.

This chapter introduces various spectra for materials that are used in paintings, such as pigments, binders, and their mixtures. The specimens used in this work are measured using a painted plate or in the form of a pellet. Figure 3.3a shows photographs of painted samples of pigments and binders on wood chips, paper, and a cyclo-olefin polymer (COP) plate. In this work, we used a commonly used commercial product, Zeonex® (Zeon Corporation, Tokyo, Japan), for the COP. It should be noted that certain private companies distribute plates or cells that are fabricated by moulding COP, and these companies then claim new names as if they had developed the material itself. COP has high transmittance in both the far-infrared region and the visible region, and is also used to fabricate THz components because of its low loss characteristics [8, 9]. Other examples of THz-transparent materials include polyethylene, polypropylene, and polymethylpentene (TPX).

Figure 3.3b shows spectra for the azurite pigment that were obtained by FT-THz (JASCO) with samples prepared either as a pellet (2 wt.% in polyethylene powder) or as a painted sample on a COP plate, as shown in the photograph in the figure. The spectra obtained from the painted sample appear dull and unclear. This could be caused by scattering at the surface, and this will be discussed in the next section. To perform a quantitative analysis, pellet samples are required. However, in practical cases, colours appear in the painted form. Therefore, most of the spectra in this chapter, unless otherwise stated, are obtained using painted samples on COP plates.

The spectra are obtained either in transmission mode or in reflection mode. Figure 3.4a–c show comparisons between the transmission and reflection spectra of cinnabar, lead white, and verdigris, respectively. The reflection spectra of cinnabar and lead white show characteristics that are almost the reverse of their respective transmission spectra; however, no theoretical investigation of these spectra has been performed to date. The reflection signal of verdigris is beneath the noise level over the entire frequency range of the system. In this chapter, the transmission spectra are mainly discussed, except for the case of spectroscopic imaging, which uses reflected waves.

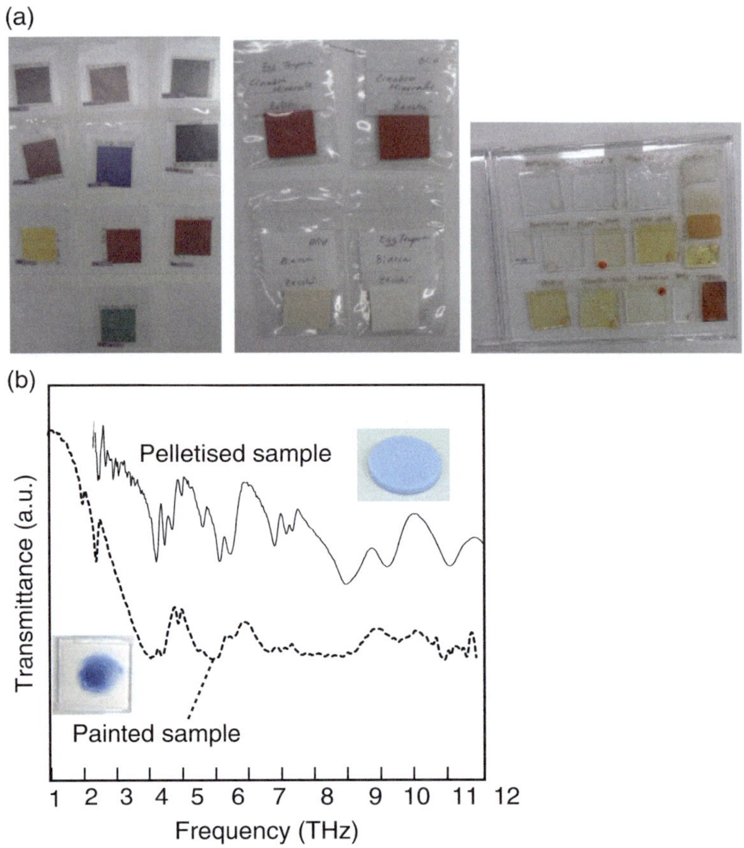

Fig. 3.3 Sample preparation for spectroscopy. (**a**) Examples of painted samples on paper, wood, and Zeonex®, (**b**) comparison of transmission spectra of painted and pelletised samples of azurite observed using the FTIR system with mercury lamp source

The consistency of spectroscopy by FTIR and TDS was confirmed in Fig. 3.5, which compares the spectra for cinnabar that were obtained by TDS and FTIR. The positions of the peaks are in good agreement, and thus THz TDS can indicate the absorption peaks properly using pulses. Here, the spectra shown in Fig. 3.5b show good agreement with the spectra that were obtained in 1970, as shown in Fig. 3.1. Because a wider frequency band is better for these measurements, FTIR was used to analyse materials in detail.

Some analytical works on spectra suggest that the frequency range in the THz region is suitable for organic-inorganic composites, which have been actively developed in the twenty-first century, because the THz frequency range corresponds to weak bonds such as hydrogen bonds, according to fundamental physics research [10–14]. THz spectroscopy may then be used by chemists in industrial applications in the future. Then, THz spectroscopy, or the "far-infrared option of FTIR", may hopefully find a larger market.

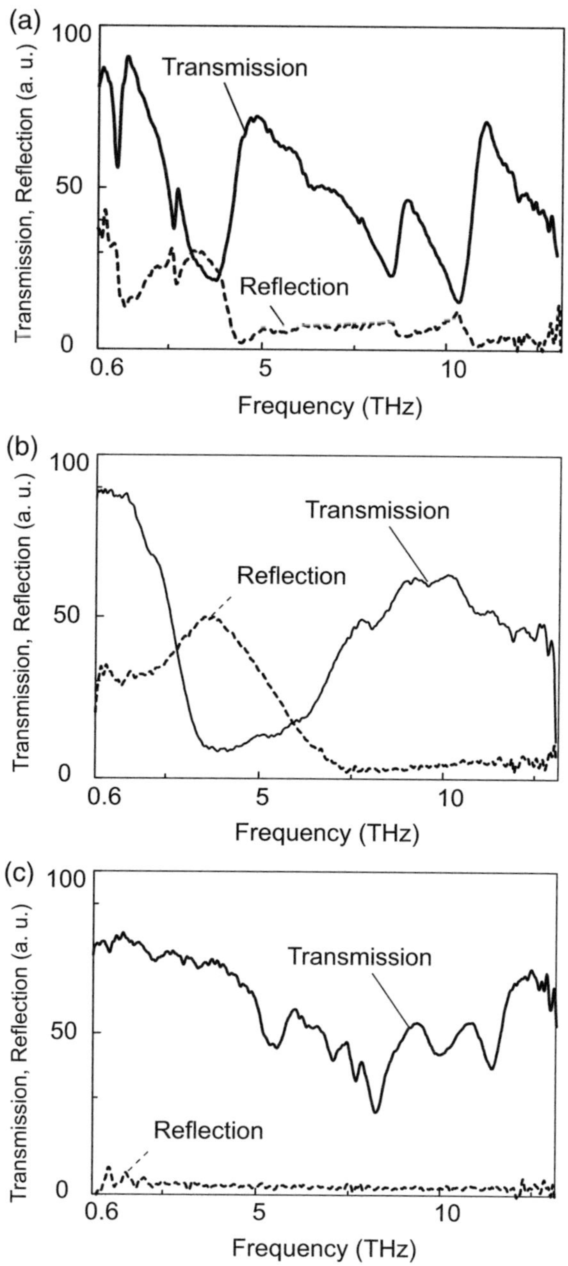

Fig. 3.4 Comparison of transmission and reflection spectra of mineral pigments using a painted specimen with painting oil that has no particular peak in the frequency range used in this study. (**a**) Cinnabar, (**b**) lead white, and (**c**) verdigris. For quantitative investigations, pelletised specimens are required

It should be noted here that the wavelengths of THz waves are close to typical pigment particle sizes, especially in Japanese paintings. As shown in Fig. 3.6a, Japanese painting methods use different particle sizes to change the colour tone.

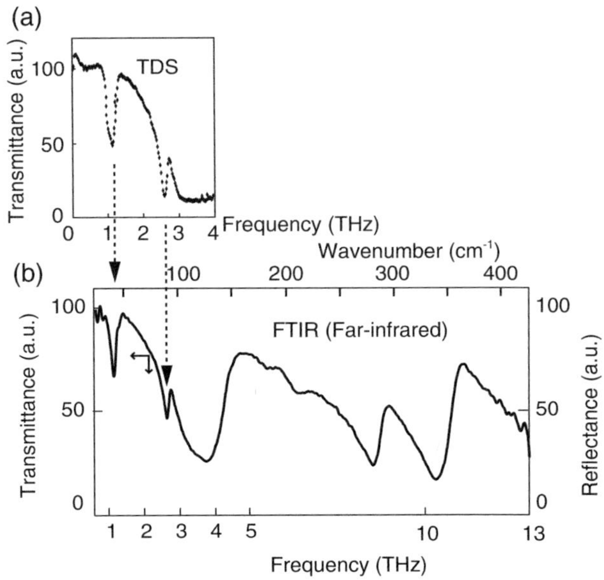

Fig. 3.5 Comparison of spectra for cinnabar obtained using TDS and FTIR. The spectra show good agreement with those shown in Fig. 3.1

The blue paints in the figure are made from the same mineral pigment shown in the photograph of the microscope observation. The differences in the visible colours are made only by changing the particle size. The particles used for pale shades of blue and green are 5 μm in diameter on average, and those used for the darker versions of the colours are 30 μm in diameter. The photograph on the left-hand side is an example from a masterpiece panel screen, "Irises at Yatsuhashi", in which the flowers and leaves were painted on gold by varying the particle sizes from extremely fine to coarse. To analyse artworks of this type, the influence of the particle size on the spectra must be considered. Figure 3.6b, c shows the spectra of azurite pellet specimens with various particle sizes, as measured by THz-TDS and FTIR, respectively. It is clear that the spectra become sharper when the pigment size is less than 40 μm. The dull features obtained when using coarse pigments could be caused by Mie scattering [15, 16].

3.3 Spectra of Art Materials

This section introduces the spectra of pigments and binders that were obtained using the FT-THz system [17]. The samples were prepared by painting on COP plates with painting oil. The oil was chosen because its transmission in the THz frequency range is almost 100 %. However, pellet-type samples are required to perform quantitative analysis.

Because the absorption peaks that appear in the THz region have not yet been assigned to the specific physical behaviour of substances, such as the phonon

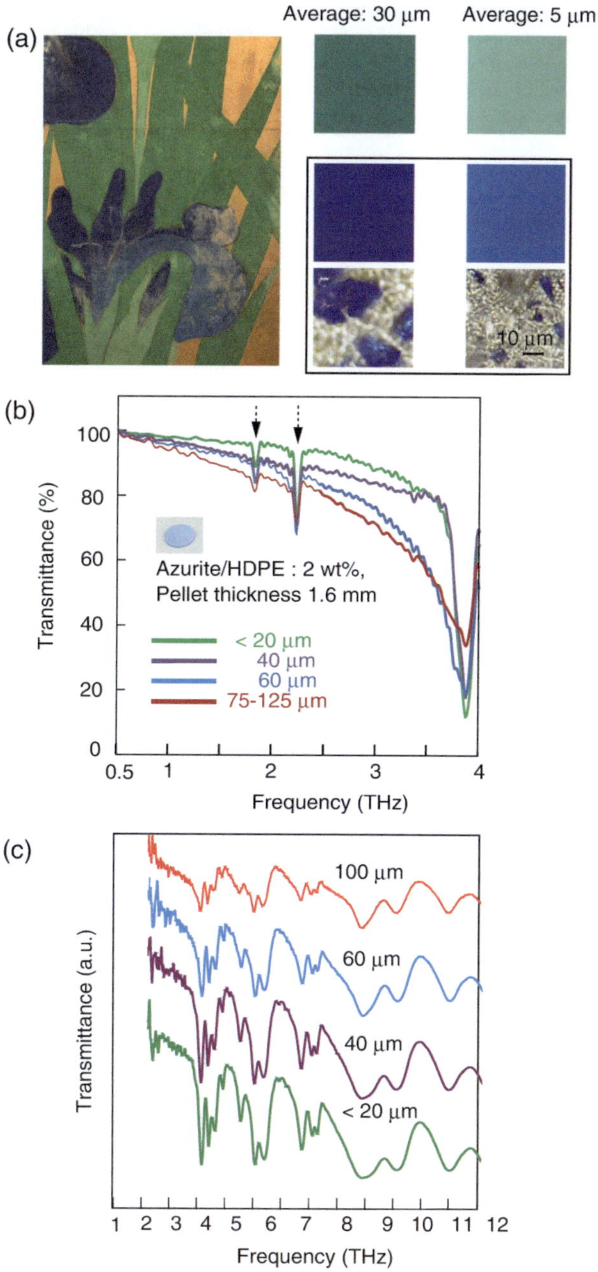

Fig. 3.6 Influence of pigment particle size on THz spectra. (**a**) Change of colours by varying particle size, and an example in the masterpiece "Irises at Yatsuhashi", by Ogata Kōrin. This technique has been commonly used in Japanese paintings to date, (**b**) spectra for azurite obtained using pelletised samples that were observed via TDS, (**c**) spectra obtained using painted samples that were observed by FTIR. These spectra show that finer particle sizes produce better spectra with clear peaks. When the particle size is as large as the wavelength of the THz wave, then the spectrum becomes broad because of Mie scattering

absorption processes discussed in a previous work [5], spectroscopy in the THz region cannot be used to "identify" materials. However, the differences between spectra may help to distinguish pigments of the same colour. Figure 3.7 shows the

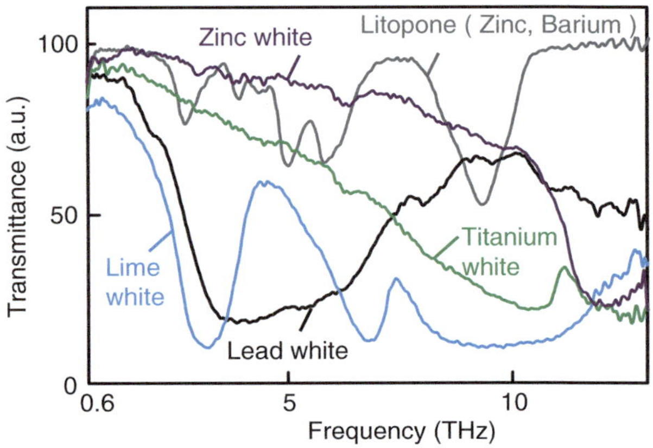

Fig. 3.7 Spectra of white pigments and an example of a mixture. The pigments under test are lime white, lead white, zinc white, lithopone (a mixture of zinc and barium), and titanium white

spectra of five white pigments that all look the same to the human eye. The transmission characteristics of each of these pigments are completely different from the others, and this suggests the possibility of performing principal components analysis (PCA) to distinguish between the materials. For example, lime white is widely used as a body pigment because it becomes transparent in paint when mixed with oil-based binders. While the presence of lime white is not visibly recognisable, its specific features can be detected easily in the THz region.

Figure 3.8 shows the spectra of various mineral pigments. Sharp peaks appear for most of the pigments, except for the carbon black and earth pigments. The spectral features are not solely determined by the presence of the main element: the molecular structure also plays an important role. Both lead white and litharge (Fig. 3.8a) show broad peaks, while minium and Naples yellow have many sharp peaks. Malachite and azurite pigments from mineral rocks have similar spectra (Fig. 3.3b). The transmission through these samples is low, and this is probably because of their relatively large particle sizes. Verderame is the same substance as malachite in terms of chemical formula, and is artificially made by the oxidation of copper. The particle size of the artificial pigment is generally much smaller than that of malachite made from natural rocks. Therefore, Mie scattering does not appear in the case of verderame. As shown in Fig. 3.7, lime white has a specific transmission spectrum, and coral red, which is also composed of $CaCO_3$, shows very similar spectra (Fig. 3.8c). Earth colour pigments, which are mostly composites of iron oxide, tend to show broad features. These earth pigments are easily obtained from natural soil, and the colour depends on the area in which they were mined (Fig. 3.8d). Black pigments based on carbon black are almost transparent in the THz region (Fig. 3.8e). However, ivory black has different features. This is because ivory black contains only a small amount of carbon black itself, and the composition is dominated by more than 80 % calcium phosphate content.

Cobalt composites, which are used in artificial mineral pigments, show sharp peaks. In the case of cobalt blue, for example, the features can be summarised into two types among 14 pigments from seven suppliers, as shown in Fig. 3.8f. Earth

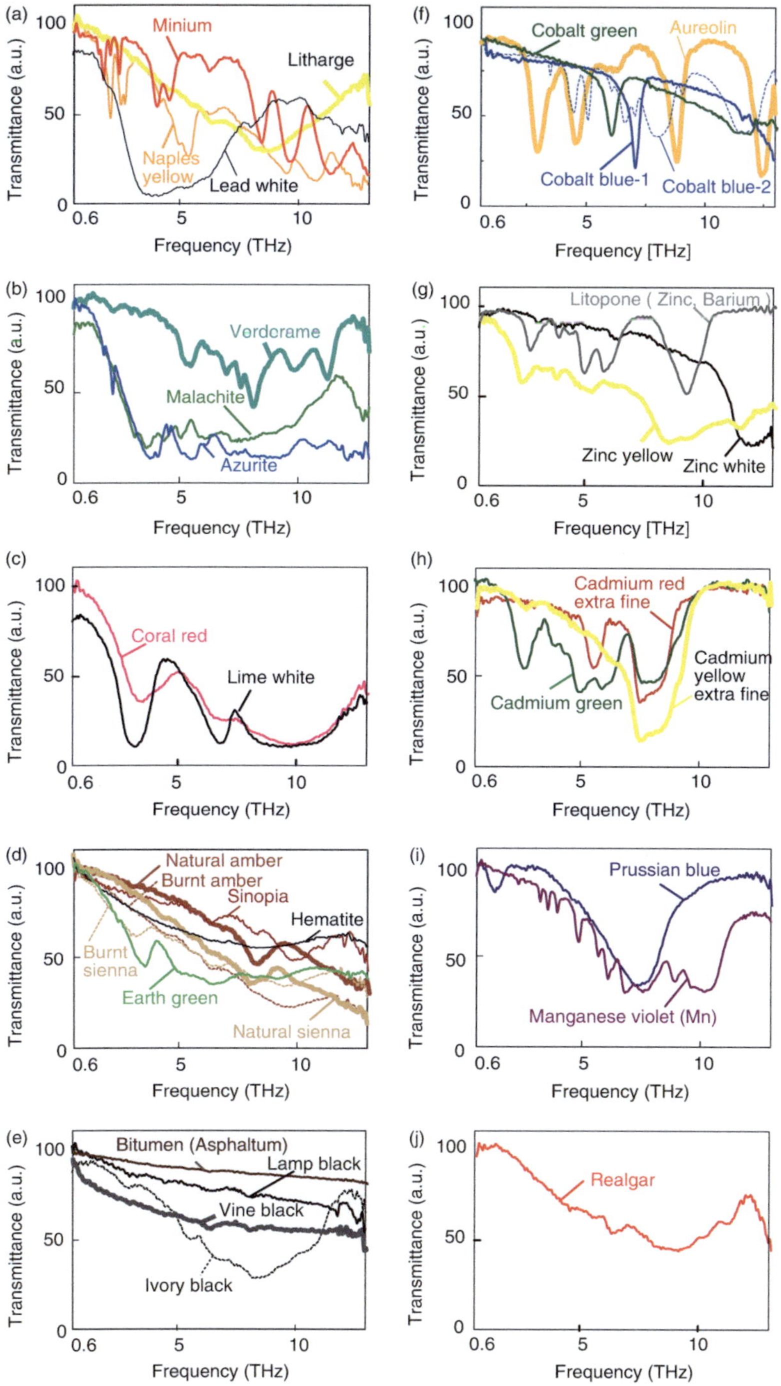

Fig. 3.8 Examples of spectra of mineral pigments. (**a**) Natural/artificial lead (*Pb*) composites, (**b**) natural/artificial copper (*Cu*) composites, (**c**) natural calcium (*Ca*) composites, (**d**) natural earth pigment (*Fe*, *Mn*, *Al*)-based composites, (**e**) natural carbon black and ivory black pigments, (**f**)

colour pigments, which are mostly composites of iron oxide, as stated above, tend to show broad absorption characteristics.

While zinc oxide shows a broad spectrum above 10 THz when it is used as zinc white, other zinc composites such as zinc yellow show several specific peaks in the lower frequency region (Fig. 3.8g). Lithopone is a mixture of $BaSO_4$ and ZnS, and zinc yellow is a composite of zinc and chrome. As demonstrated in this example, it is clear that the principal element does not determine the overall spectrum simply by its presence.

Cadmium yellow is CdS, and shows a specific relatively broad peak at around 7 THz (Fig. 3.8h). Cadmium red and orange are made from mixtures of CdS and CdSe, and a new peak around 5.1 THz appears for these materials. Cadmium green is a mixture of viridian and cadmium yellow. However, the features of this mixture are similar to those of lithopone that were shown in Fig. 3.8g. Because $BaSO_4$, which is a component of lithopone, is also used as a body pigment, peaks appear both in Fig. 3.8g, h.

Prussian blue is an artificial colour containing iron ($C_{18}Fe_7N_{18}$). Unlike iron oxide, a broad peak appears for this material in the THz region. This may be because the iron in Prussian blue exists as a complex salt. Manganese violet ($H_4NMnO_7P_2$) shows small sharp peaks over a broad absorption range (Fig. 3.8i).

Some natural mineral pigments are on the verge of extinction. As one example, realgar, for which spectra are shown in Fig. 3.8j, is an arsenic composite. To perform research on historic paintings, it is important to have records of material properties using standard samples so that we actually know the correct chemical formulas. Therefore, samples of pigments that are not ecologically friendly should be conserved for the next generation of researchers. It should be noted that the current tendency to ban every single material that contains lead, cadmium or any substance that causes environmental problems may badly affect the heritage science research field.

Figure 3.9 shows natural and artificial organic dyes. Cochineal, which is extracted from insects, and sepia are almost transparent, as shown in Fig. 3.9a. The spectra of dyes that originate from plants tend to vary (Fig. 3.9b). Dyes that are derived from roots tend to exhibit sharp peaks. Here, the spectra of gal inks, which should contain iron, show similarities to the spectra of earth pigments. Artificial dyes are commonly used in modern paintings and the amounts and types of these dyes will increase continuously. Figure 3.9c shows the spectra of rather traditional artificial dyes. Clear sharp peaks appear in the THz region. These materials are developed in industry, and thus this suggests that there is a possibility of performing spectral assignment in the future using the pure designed specimens.

In general, the material identification of pigments can be performed sufficiently using existing nondestructive methods such as XRF, XRD, and Raman spectroscopy. For the analysis of varnishes and binders, however, sampling is

---

Fig. 3.8 (continued) artificial cobalt (*Co*) composites, (**g**) artificial zinc (*Zn*) composites, (**h**) artificial cadmium (*Cd*) composites, (**i**) artificial iron (*Fe*) and manganese (*Mn*) composites, (**j**) natural arsenic (*As*) pigment

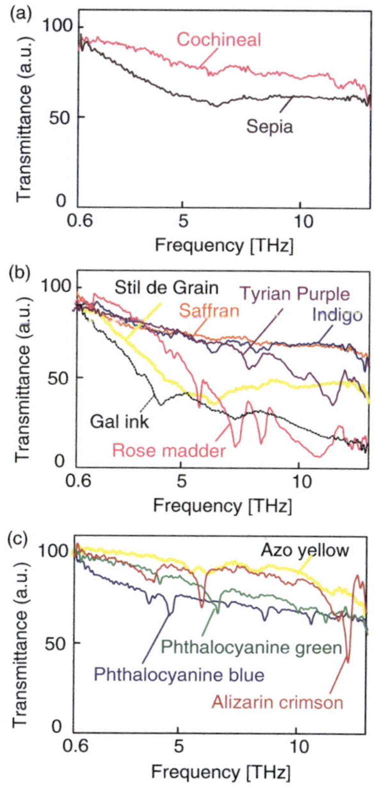

Fig. 3.9 Examples of spectra of organic dyes. (**a**) Natural dyes derived from insects and squid, (**b**) natural dyes derived from plants, (**c**) artificial dyes

required to perform analyses via mid-infrared spectroscopy or other chemical analysis methods. Feasibility studies of nondestructive analysis using THz spectroscopy were performed by several institutions to determine whether specific peaks from these binders appear in the THz region or not. However, experiments using FTIR have produced negative results to date, as shown in Fig. 3.10. Another work using THz TDS [18] shows similar tendencies. Most binders, both natural and artificial, are relatively transparent in the THz frequency region (except for beeswax, as shown in Fig. 3.10c), and thus it is extremely difficult to identify binders in paintings using THz spectroscopy. From an imaging perspective, however, the transparency of varnish in the THz region is an advantage because the THz waves can penetrate into the painting layer, and unlike the mid-infrared range, the spectra of the pigments are not affected by the presence of the varnish.

Here, a comparison of the spectra revealed that mixed binders have spectral features that are derived from the mixed substances. For example, Venetian turpentine is a natural mixture of oil and resin, as confirmed by the results shown in Fig. 3.11a. Beva® is a commonly used material that was developed for conservation purposes, and is based on a mixture of polyvinyl acetate (PVAc) and natural resin. The spectrum of the Beva® film shown in Fig. 3.11b shows broad peaks from

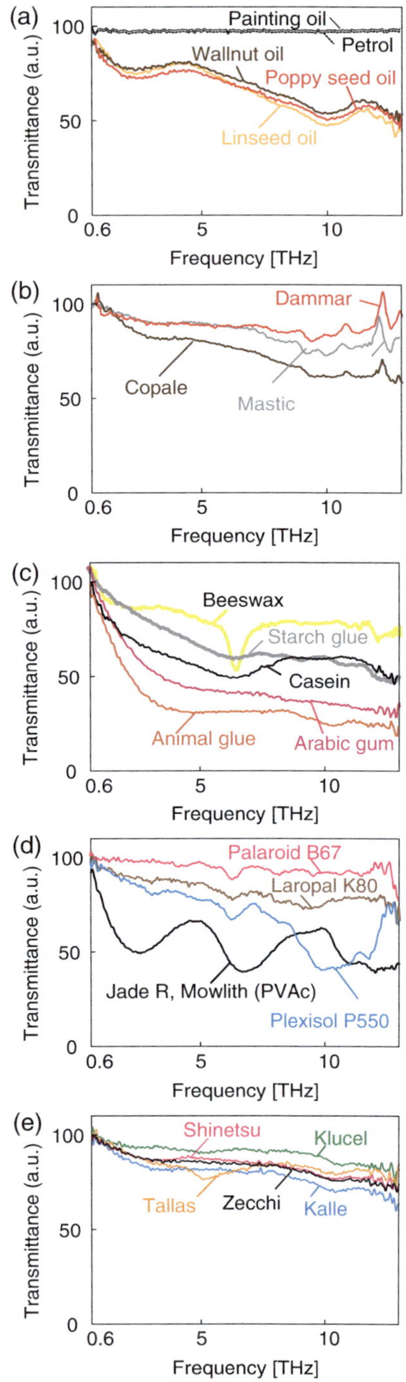

Fig. 3.10 Examples of spectra of binders. (**a**) Oils, (**b**) natural resins, (**c**) various binders, beeswax, casein, gum arabic, and glues, (**d**) artificial polymeric binders, (**e**) artificial cellulose-type glues

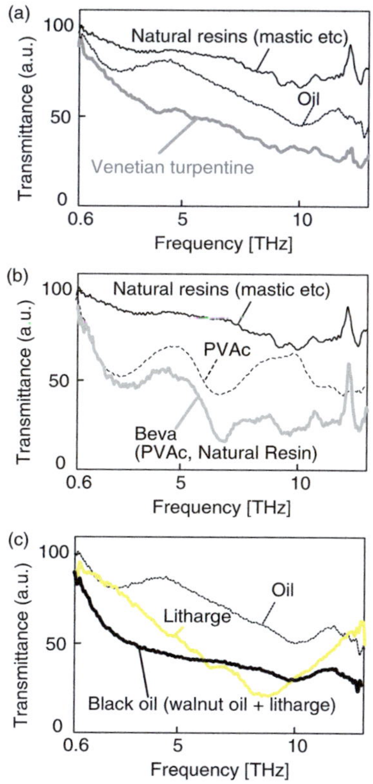

Fig. 3.11 Examples of spectra of mixed binders. (**a**) Venetian turpentine or oleo-resin: a natural mixture of oil and resin. The spectrum has features from both the oil and the resin, (**b**) Beva: one of a series of products developed for conservation, based on a mixture of PVAc and natural resin. The spectrum shows the peaks as if the two spectra were superimposed, (**c**) black oil: a mixture of walnut oil and litharge. The influence of litharge can be recognised in the spectrum of black oil, when compared with the spectra of natural oils

PVAc (Fig. 3.10d) and from the kind of resin used (Fig. 3.10b). Black oil is an artificial quick-dry painting oil, and is made by mixing litharge (Fig. 3.8a) with natural oil (Fig. 3.10a). The influence of litharge can be recognised in the spectrum of black oil when compared with the spectra of natural oils.

In the case of paints, which are mixtures of pigments and binders, spectral features from the two substances appear as shown in Fig. 3.12. A type of cobalt blue pigment was used to make paint with a mixture of natural oil (Fig. 3.12a) and acrylic resin (Fig. 3.12b). The peak from the cobalt blue in this work is sharp enough to be distinguished from the broad absorption characteristics of both binders.

While Figs. 3.11 and 3.12 can be read as mixtures of these spectra, the spectrum of black oil cannot directly suggest the two spectra of litharge and oil. In this way, we see that while the use of THz technology for spectroscopy seems to be useful for material investigations, it is still too optimistic to suggest that the technology can be used for substance identification at this stage.

In laboratory experiments, THz spectroscopy can provide useful information for the examination of pigments. Figure 3.13 shows the spectra of cobalt blue and cerulean-related pigments that are commercially available in Japan. There are two

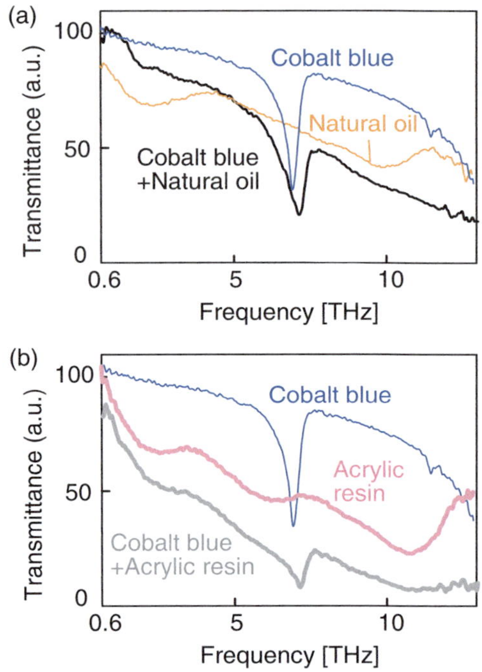

Fig. 3.12 Examples of spectra of paints when using a type of cobalt blue pigment. (**a**) Mixture with natural oil, where the sharp peak of the cobalt blue pigment was affected by the presence of the oil, (**b**) mixture with a type of acrylic resin. The spectra clearly show that the spectrum of each mixture is formed by the spectra of both components, and no new peaks appeared

types of spectral features in the cobalt blue pigments, as shown in Fig. 3.13a, b. The relatively broad spectra that appeared in Fig. 3.13a were obtained using inexpensive "standard" grade pigments. This tendency is discussed later with reference to Fig. 3.15. All the cerulean blue pigments show the same features, and these features are totally different from the two spectra of the cobalt blue pigments. Among the specimens that are available in Japan, one cobalt blue pigment has the same spectrum as the cerulean blue pigments.

However, the names of paints are not strictly determined by their chemical components, and artists choose colours depending on their own eyes. Therefore, this could become a problem only for who try to match the specimens' names with the specific substances.

Figure 3.14 shows the spectra of ultramarine-related pigments. The particle sizes in each sample were smaller than 10 μm to avoid any factors that would affect the apparent absorption, as shown earlier in Fig. 3.6. Artificial pigments, in general, have higher THz wave transmittance properties than natural pigments. This could be due to the presence of crystalline water, which exists in natural rocks. Figure 3.14b shows the spectra of two pigments that are believed to be produced based on artificial ultramarine. The ultramarine rose shows additional peaks around 8 THz. The cobalt blue light pigment from one company did not show the two specific spectra of cobalt blue that are shown in Fig. 3.13a, b. Based only on these

Fig. 3.13 Spectra of cobalt blue and cerulean-related pigments that are available in Japan. (**a**, **b**) Cobalt blue pigments, (**c**) cerulean blue pigments, (**d**) a pigment that is sold as cobalt blue light, but contains cerulean blue

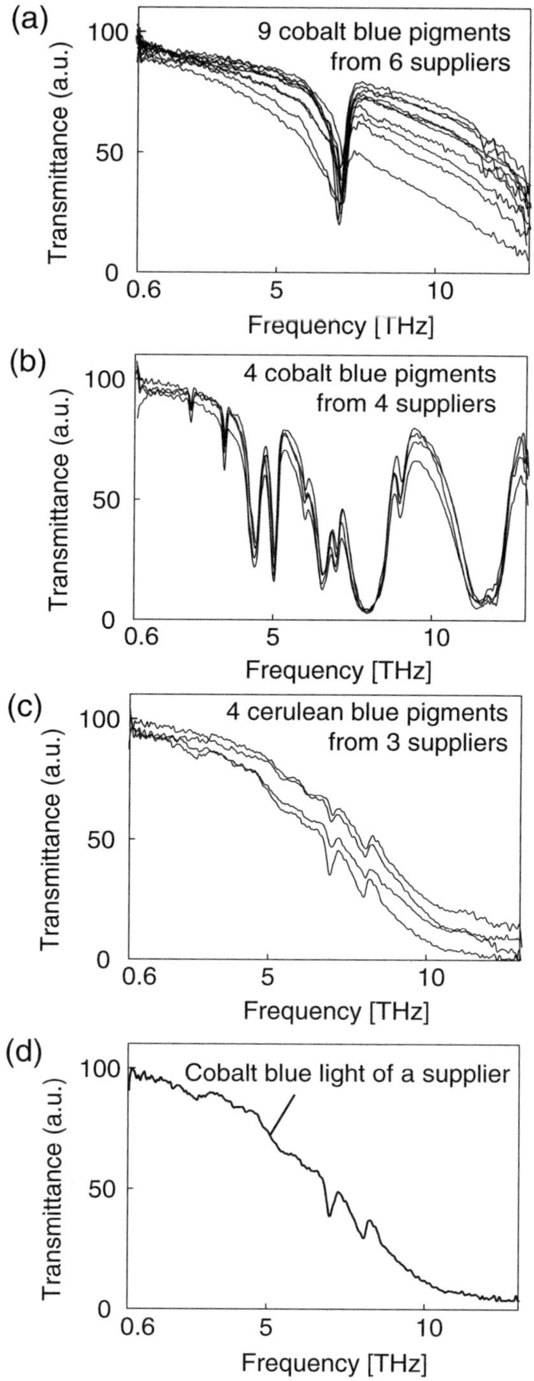

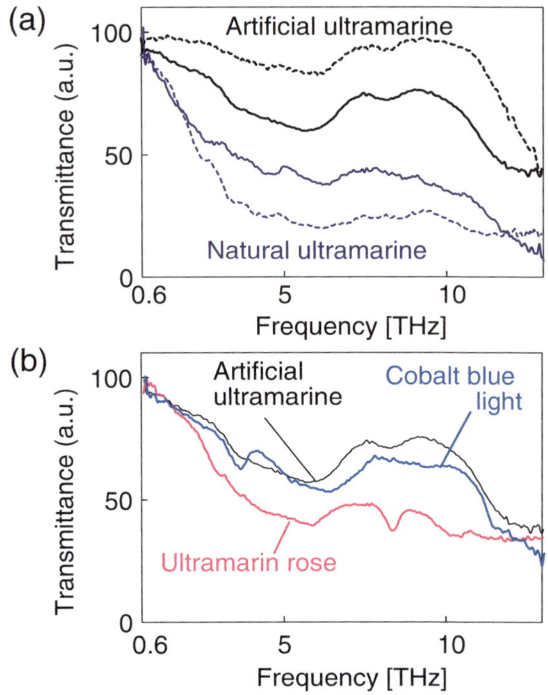

Fig. 3.14 Spectra of ultramarine related pigments. (**a**) Comparison of natural and artificial pigments, (**b**) pigments that are likely to be made from artificial ultramarine

THz spectra, this particular cobalt blue seems to have been made using artificial ultramarine, and contains some features of earth green.

Figure 3.15 indicates that THz spectroscopy can be useful for detecting filler components and/or contamination of pigments or to provide some hints to aid in substance identification. Figure 3.15a shows that the spectra of pigments depend on their purity, in the case of cadmium red. The spectrum for "extra fine" shows clear spectra for the components, while "standard" shows broader features. These differences can be caused by impurities or substances that have been added on purpose, such as body pigments. A body pigment such as lime white can be added to other colours to increase the overall volume because it becomes transparent when mixed with oil-based binders. Figure 3.15b shows the spectra of two emerald green pigments. The genuine emerald green pigment shows little absorption, and has tiny peaks. The standard emerald green pigment, in contrast, has the same spectrum as that of lime white. This proves that the standard paint contains a certain amount of lime white as a body pigment. The term "standard", when used for paints, often means nonspecialised affordable materials, and thus it is understandable that standard pigments contain body pigments. Although the colours look similar, the spectra in the THz band reveal the true quality of the paints. When it comes to materials identification, classic laboratory testing methods are available, and THz spectroscopy may thus not be useful in a practical sense. Figure 3.15c shows the

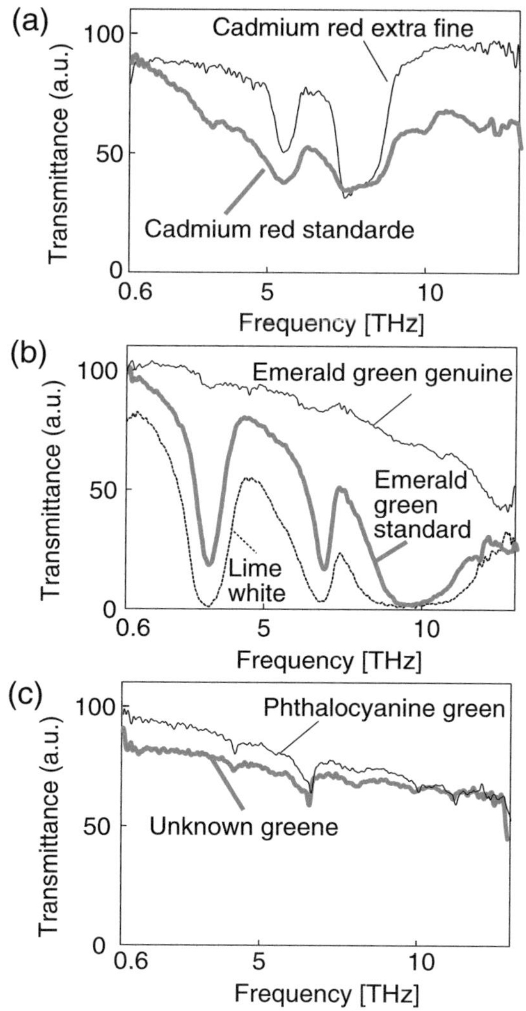

Fig. 3.15 Spectra that can be used for materials investigation in practice. (**a**) Two cadmium red pigments, extra fine and standard, (**b**) two emerald green pigments, genuine and standard, compared with the spectra of lime white, (**c**) phthalocyanine-type pigments

spectrum of an unknown green obtained from a shop for conservators, and the spectrum is close to that of phthalocyanine green. However, this information is not sufficient for substance identification.

Peak assignment for material identification has not yet been an active research field in THz region. This is quite reasonable because the information that can be extracted from these broad peaks can be very limited.

3.4 Spectroscopic Imaging

The spectroscopic information of pigments is likely to generate interest in spectroscopic imaging. Figure 3.16 shows an experimental THz-TDS imaging trial of a model sample that was painted using pigments with large differences in transmission on a COP plate. Artificial and natural ultramarine pigments were used for the blue paints, while zinc white and lead white were used to paint the backgrounds. Cadmium red and cinnabar are used as the red colours. All common black pigments are transparent in the THz region, and thus PVAc was chosen as a binder to obtain the transmission images. Other colours were painted using common painting oil, which is transparent in the THz frequency region, as shown in Fig. 3.10. Figure 3.16a–c shows the spectra of the materials used in the imaging experiments, and the results of these experiments are shown in Fig. 3.17. Figure 3.17a shows a red half circle that contains a flower, painted with cadmium red, surrounded by a red background painted with cinnabar. The flower is hardly visible to the naked eye at a distance, while THz transmission images clearly show the image, although the background cinnabar painted area was not homogeneous because of the rough brush

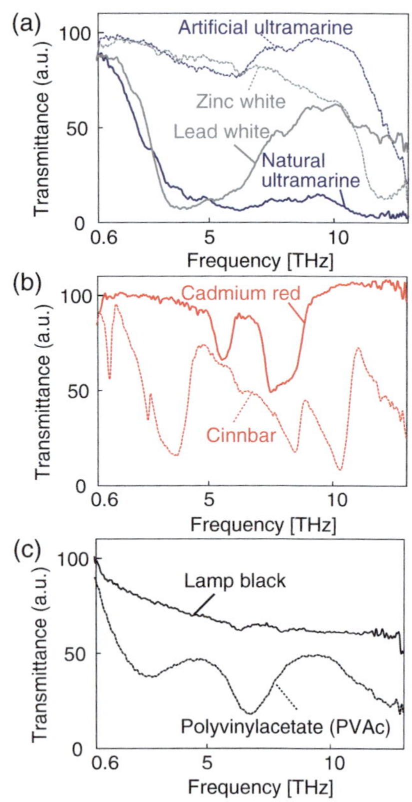

Fig. 3.16 Possibilities of spectroscopic imaging. (**a**) Spectra of two *white* and *blue* pigments, (**b**) spectra of two *red* pigments, (**c**) spectra of *black* pigment and two binders

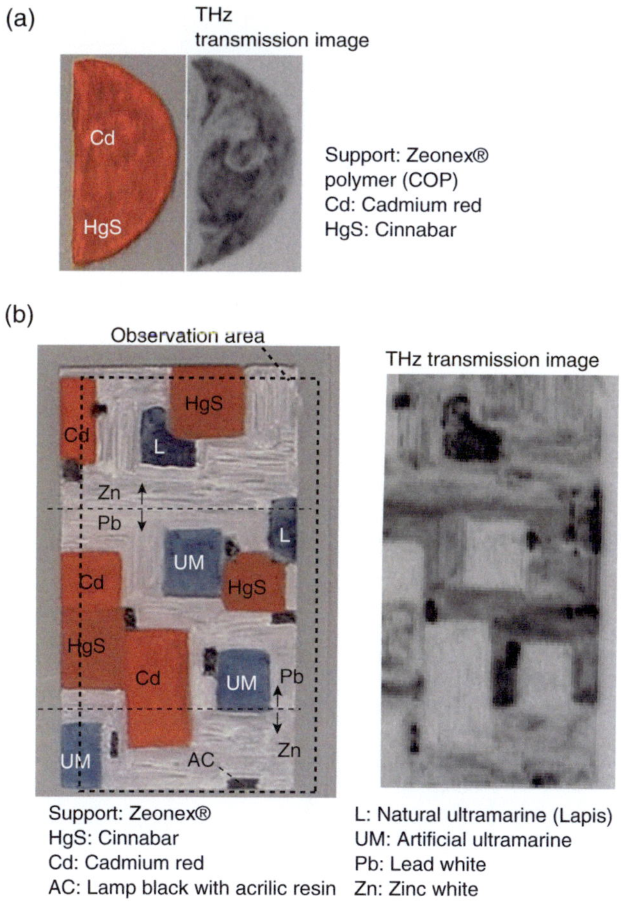

Fig. 3.17 (**a**) Transmission imaging of *red* painted model on a COP plate, (**b**) transmission imaging of a multi-coloured painted model on a COP plate

strokes. Figure 3.17b shows a model with four colours. As shown in the THz transmission image, the difference between the two parts painted with ultramarine is large enough to be recognisable. The central part of the background white was painted using lead white, which resulted in relatively dark features. The black paint containing PVAc exhibits the darkest spots. These model experiments suggest that spectroscopic imaging is possible if the properties of the materials are well known.

As a trial based on the use of a real specimen from the past, transmission imaging was applied to a parchment sample from the thirteenth century, as shown in Fig. 3.18a. First, a tiny amount of red paint was sampled to identify the materials in transmission mode. It is well known that the red pigments used in the red inks in medieval parchments were minium (a lead composite) and cinnabar. As the comparison in Fig. 3.18b shows, several peaks that appeared in the spectrum of the

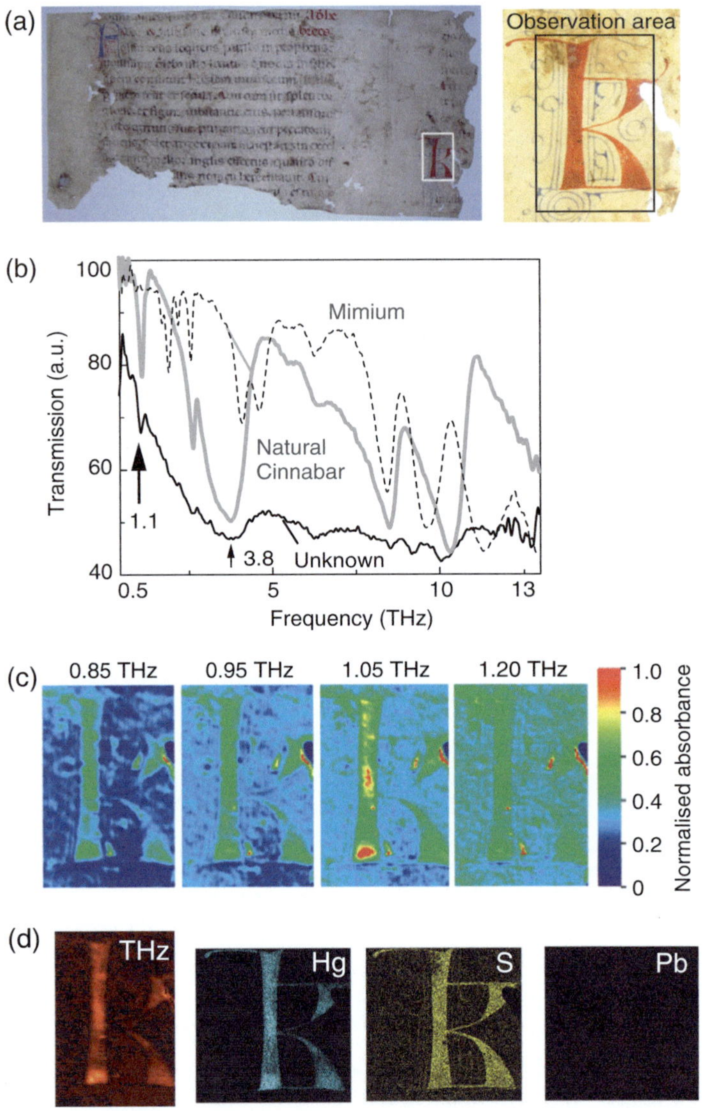

Fig. 3.18 THz transmission imaging example of parchment sample from thirteenth century. (**a**) Photograph of the sample and an enlarged image of the observation area, (**b**) spectrum of unknown *red ink* from the sample, containing natural cinnabar and minium, (**c**) multi-spectral THz transmission images used in principal component analysis, (**d**) extracted image of the letter K and comparison with XRF mapping of the same observation area

unknown red colour are well matched to those in the cinnabar spectrum. Multi-spectral THz transmission images were then prepared for principal component analysis, as shown in Fig. 3.18c. Finally, the letter K is extracted as shown from the image in Fig. 3.18d. The THz image was compared with that produced by XRF

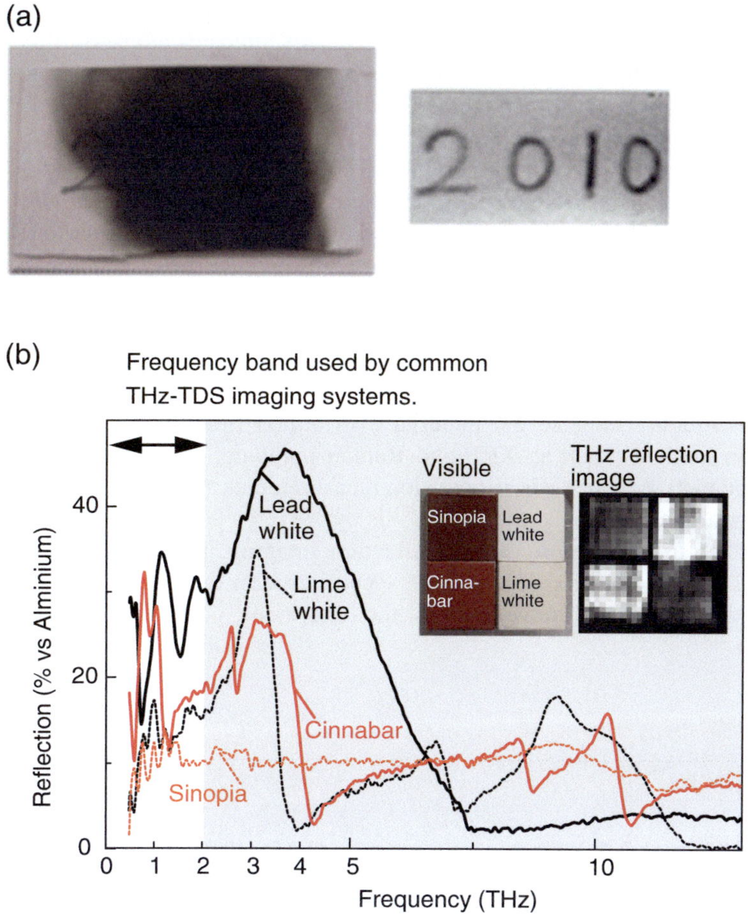

Fig. 3.19 Examples of THz reflection imaging. (**a**) Graphite drawings of "2010" under burned carbon black, (**b**) painted wood pieces containing sinopia, cinnabar, lead white, and lime white. The binder is egg, as used in tempera painting

mapping. The XRF mapping results proved that the ink is based on cinnabar (HgS) and not on minium (Pb_3O_4).

While this experiment may give some impression that THz pulsed TDI can be of practical use for text recognition, it was performed in transmission mode, and thus a trial observation in reflection mode must also be performed [19]. Figure 3.19 shows an example of the reflection imaging of the number "2010" written on the sample using a graphite pencil, before the surface was burnt. The number "2010" is covered with carbon black and is invisible to the naked eye, except for a tiny part of the "2". Because THz waves can pass through carbon black after it has been burnt, and because graphite reflects THz waves, the number "2010" can be clearly recognised. Historic writings on paper are usually made using carbon black, and graphite pencils were only invented in the mid-sixteenth century. When a drawing under paint becomes visible using THz pulsed TDI, this indicates the possibility that the artwork was painted after the development of graphite pencils. Useful model experiments with regard to drawings, including those on multi-layered papyrus, have been performed [20, 21].

Figure 3.19b illustrates the potential of THz pulsed TDI for use in distinguishing pigments of the same colour. The specimens were painted on wood chips, using sinopia and cinnabar for the red paints, and lead white and lime white for the white paints. Both sets of specimens have large differences between their reflection levels, and these differences are especially obvious in the frequency range from 0.5 to 2 THz, which is used for commercially available THz-TDS imaging systems. The grey scale description of the THz wave reflection level makes it possible to display the difference, resulting in production of a substance map. THz pulsed TDI can thus roughly indicate the material distribution in a shorter time than other precision methods such as XRF and Raman mapping. In other words, the THz reflection map indicates where precision measurements should be performed using other instruments, e.g., elemental analysis by XRF.

Whether using transmission or reflection imaging, it is important to extract material information from each of the spectra used in the imaging process. Some recent works [22–24] have proved the enormous potential of spectroscopic imaging.

References

1. R.S. McDonald, P.A. Wilks Jr., Appl. Spectrosc. **42**, 151 (1998)
2. Infrared Raman Users' Group website, http://www.irug.org/
3. B.A. Price, B. Pretzel, S.Q. Lomax, C. Davis, J.H. Carlson, Revised JCAMP-DX spectral file format for submissions to the infrared Raman Users Group (IRUG) spectral database, available from [2]
4. E.D. Palik, *Handbook of Optical Constants of Solids* (Academic, New York, 1997)
5. C. Karr Jr., J.J. Kovach, Far-infrared spectroscopy of minerals and inorganics. Appl. Spectrosc. **23**, 219–223 (1969)
6. H.D. Riccius, K.J. Siemsen, Infrared lattice bands of trigonal and cubic mercury sulfide. J. Chem. Phys. **52**, 4090–4093 (1969)
7. R.A. Nyquist, R.O. Kagel, *Infrared Spectra of Inorganic Compounds (3800-45 cm^{-1})* (Academic, New York, 1971)
8. F. Pavanello, F. Garet, M.-B. Kuppam, E. Peytavit, M. Vanwolleghem, F. Vaurette, J.-L. Coutaz, J.-F. Lampin, Broadband ultra-low-loss mesh filters on flexible cyclic olefin copolymer films for terahertz applications. Appl. Phys. Lett. **102**, 111114 (2013)
9. Y. Kishi, M. Nagai, J.C. Young, K. Takano, M. Hangyo, T. Suzuki, Terahertz laminated-structure polarizer with high extinction ratio and transmission power. Appl. Phys. Express **8**, 032201 (2015)
10. K. Liu, M.G. Brown, J.D. Cruzan, R. Saykally, Terahertz laser spectroscopy of the water pentamer: structure and hydrogen bond rearrangement dynamics. J. Phys. Chem. A **101**, 9011–9021 (1997)

11. C.A. Schmuttenmaer, Exploring dynamics in the far-infrared with terahertz spectroscopy. Chem. Rev. **104**, 1759–1779 (2004)
12. H. Yada, M. Nagai, K. Tanaka, Origin of the fast relaxation component of water and heavy water revealed by terahertz time-domain attenuated total reflection spectroscopy. Chem. Phys. Lett. **464**, 166–170 (2008)
13. M. Takahashi, Terahertz vibrations and hydrogen-bonded networks in crystals. Crystals **4**, 74–103 (2014)
14. Y. Ohki, M. Okada, N. Fuse, K. Iwai, M. Mizuno, K. Fukunaga, Terahertz time-domain spectroscopic analysis of molecular behavior in polyamide nanocomposites. Appl. Phys. Express **1**, 122401 (2008)
15. A. Bandyopadhyay, A. Sengupta, R.B. Barat, D.E. Gary, J.F. Federici, M. Chen, D.B. Tanner, Effects of scattering on THz spectra of granular solids. Int. J. Infrared Milli. Waves **28**, 969–978 (2007)
16. C.J. Strachan, T. Rades, D.A. Newnham, K.C. Gordon, M. Pepper, P.F. Taday, Using terahertz pulsed spectroscopy to study crystallinity of pharmaceutical materials. Chem. Phys. Lett. **390**, 20–24 (2004)
17. K. Fukunaga, M. Picollo, Terahertz spectroscopy applied to the analysis of artists' materials. Appl. Phys. A **100**, 591–597 (2010)
18. J.-M. Manceau, A. Nevin, C. Fotakis, S. Tzortzakis, Terahertz time domain spectroscopy for the analysis of cultural heritage related materials. Appl. Phys. B Lasers Opt. **90**, 365–368 (2008)
19. K. Fukunaga, Y. Ogawa, S. Hayashi, I. Hosako, Application of terahertz spectroscopy for character recognition in a medieval manuscript. IEICE Electron. Expr. **5**, 223–228 (2008)
20. E. Abraham, A. Younus, A. El Fatimy, J.C. Delagnes, E. Nguéma, P. Mounaix, Broadband terahertz imaging of documents written with lead pencils. Opt. Commun. **282**, 3104–3107 (2009)
21. J. Labaune, J.B. Jackson, S. Pagès-camagna, G.A. Mourou, I.N. Duling, M. Menu, Papyrus imaging with terahertz time domain spectroscopy. Appl. Phys. A. **100**, 607–612 (2010)
22. T. Bardon, R.K. May, P.F. Taday, M. Strlic, Systematic study of terahertz time-domain spectra of historically informed black inks. Analyst **138**, 4859–4869 (2013)
23. R. Radpour, N. Bajwa, J. Garritano, S. Sung, M. Balonis-Sant, P. Tewari, W. Grundfest, I. Kakoulli, Z. Taylor, THz pulsed TDI studies of painted samples to guide cultural heritage investigations at the Enkleistra of St. Neophytos in Paphos, Cypris. Proc. SPIE. **9199**(91990Q) (2014)
24. M. Naftaly, J.F. Molloy, G.V. Lanskii, K.A. Kokh, Y.M. Andreev, Terahertz time-domain spectroscopy for textile identification. Appl. Optics **52**, 4433–4437 (2013)

Chapter 4
THz Pulsed Time-Domain Imaging (THz Pulsed TDI)

As a result of the progress made in laser technology, a system composed of a photoconductive antenna and a femtosecond fibre laser can be used to generate a picosecond-order short pulse, which can then be used as a probe to observe the internal structure of an object. The frequency range that can be obtained when using FFTs can reach the THz frequency region, and thus the pulse is called a THz pulse. Details of the generation of THz pulses and of various other types of THz wave were presented in the previous reviews and handbooks that were introduced in Chap. 2. This chapter describes how a THz pulse in the time domain can reveal the internal structures of opaque objects, and gives examples obtained using model specimens.

4.1 THz Reflection Image

Figure 4.1 shows a schematic of the principles of depth profiling using THz-pulsed TDI. When a THz pulse is applied to a sample with a double-layer structure, as shown in Fig. 4.1a, the first reflected pulse is obviously generated by the surface, i.e., it occurs at the interface between the air and the material shown in grey. THz pulses can penetrate most dielectric materials, and the second pulse is generated when the THz pulse reaches internal interface A, i.e., the surface of another material with a different refractive index. The third pulse is then generated at internal interface B and the signal then passes through to the air. As shown on the right-hand side of Fig. 4.1a, the reflection pulse sequence appears with a delay that is determined by the depth below the surface and the propagation velocity. When the material is uniform, the propagation velocity is calculated from the material's refractive index. This technique is known as the time-of-flight method, and is used in various other NDT techniques. The incident THz pulses produced by all commercial THz pulsed TDI systems are not ideal impulses, and the output signal from a single interface has two polarities. It thus appears as if there were two

K. Fukunaga, *THz Technology Applied to Cultural Heritage in Practice*, Cultural Heritage Science, DOI 10.1007/978-4-431-55885-9_4

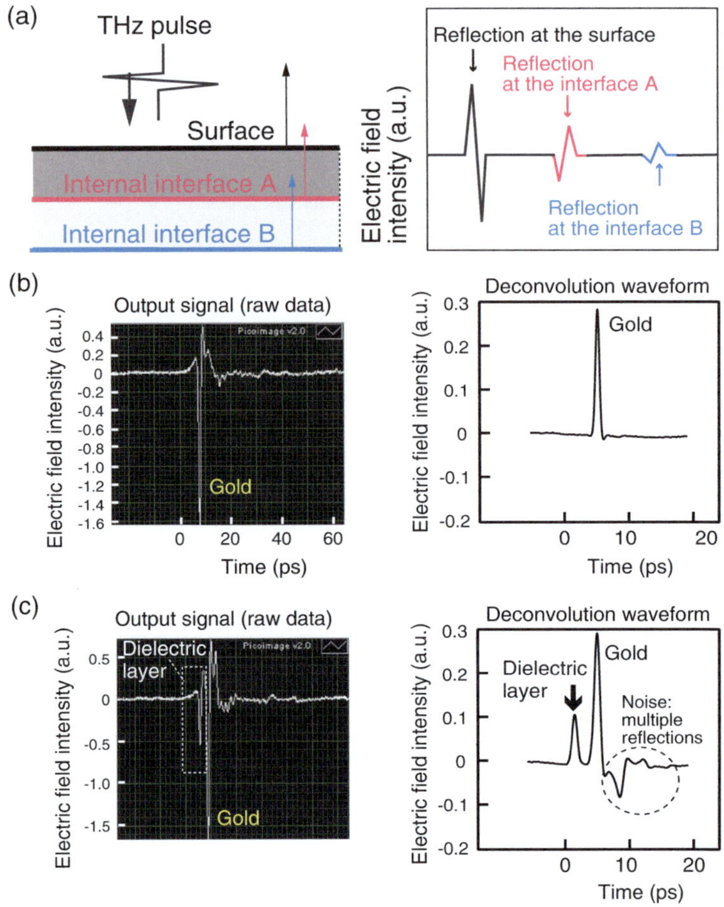

Fig. 4.1 Time domain imaging and deconvolution technique. (**a**) Schematic diagram of depth profiling by time domain imaging, (**b**) reflection from metal surface as reference signal for deconvolution, including output signal and deconvolution waveform, (**c**) example of reflection signal from dielectric layer on metal surface, including output signal and deconvolution waveform

interfaces that generated a positive pulse and a negative pulse independently. To prevent misunderstanding of this output signal, the deconvolution technique is required [1, 2]. Figure 4.1b shows an output signal from a metal surface (e.g., a gold surface) and its deconvolution signal. The transfer function for deconvolution is obtained by assuming that the reflection signal from the metal surface is a single impulse signal. Figure 4.1c shows an example of the effect of deconvolution when applied to observation of a paint layer on gold leaf. The reflection from the paint, which acts as a dielectric layer on the metal, can be distinguished. In this case, multiple reflections occur inside the paint layer between air and gold. This type of noise may produce misleading experimental data.

By scanning the object in a line with the sensing unit, which is composed of a source, a mirror and a detector, a cross-sectional image along that line is obtained. As shown in Fig. 4.2, each waveform of the output signal, e.g., the waveform at

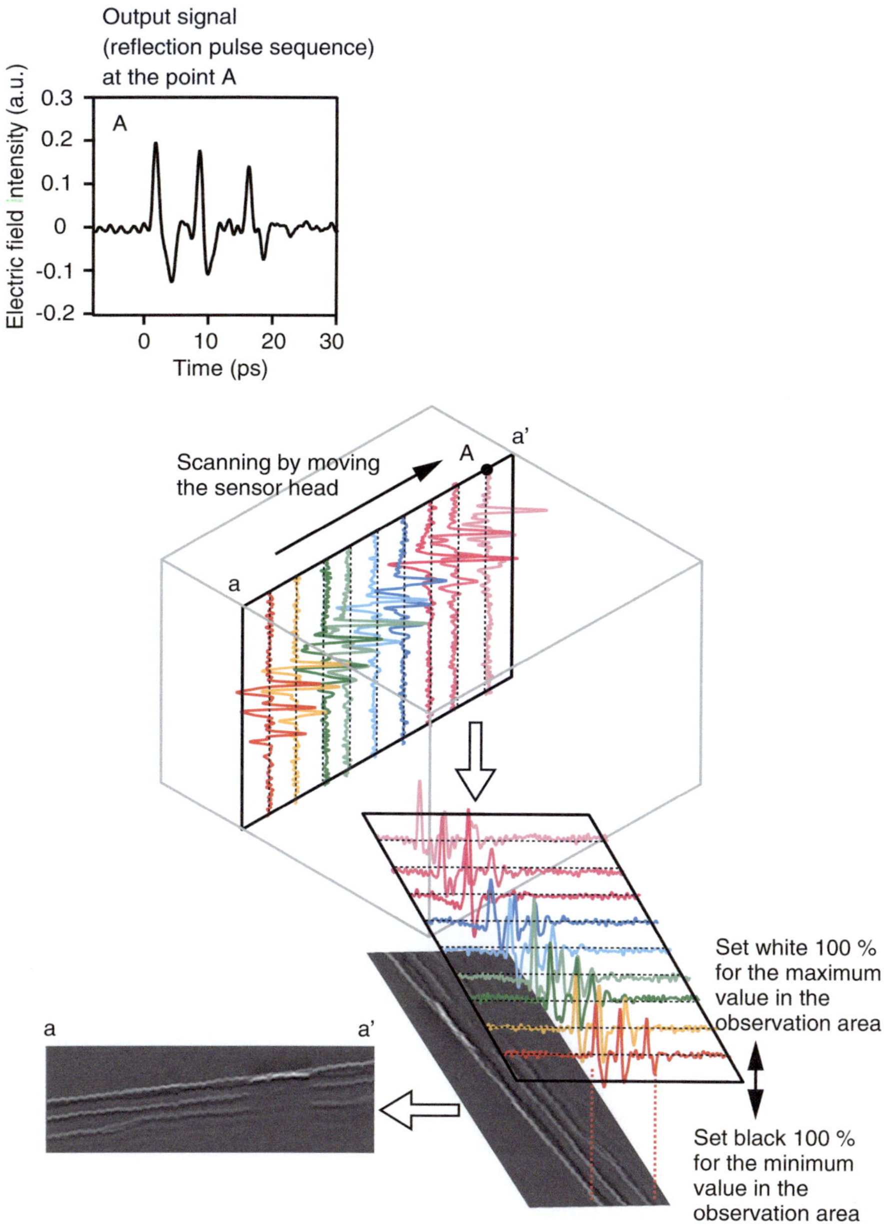

Fig. 4.2 Schematic depiction of cross-sectional image formation

point A, shows the depth profile at that point as a pulse sequence. By moving the sensing unit along the line denoted by a-a′ using a scanning system, a set of depth profiles can be obtained. Then, by setting the maximum value to be white and the minimum value to be black, a cross-sectional image along the line a-a′ is obtained.

An area image of the sample can be obtained by placing values into grey scale at each measurement point. These values can be calculated by integrating the entire signal, as shown in Fig. 4.3a, or a particular reflection pulse, as shown in Fig. 4.3b. By extracting a specific reflection pulse for the observation area, the condition of the layer of interest, such as a preparation layer for a painting, can be examined. Figure 4.4 shows an example of an area image observation using the case of the "Cypress Tree", a masterpiece by Kano Eitoku. Figure 4.4a shows the THz pulsed TDI area image produced by integrating the entire signal. Pale coloured areas indicate the gold leaf, and its high reflection value covers the details of the painting. Figure 4.4b shows the image that was extracted from the preparation layer, where pieces of mending paper of various shapes were observed by avoiding the areas of strong surface reflection. The details of the examination shown in this artwork are discussed in more detail in Sect. 6.5.

4.2 THz Pulsed TDI of Model Samples

The following contains examples of the application of THz pulsed TDI to various model samples to provide an overview of the technique.

Figure 4.5 shows the observation of internal defects in a plastic plate. The sample shown in Fig. 4.5a contains three types of internal defects, which are shown in Fig. 4.5b. THz pulsed TDI revealed all defects that corresponded to the three given types, as shown in Fig. 4.5c. For example, a double-layered signal and additional multiple reflection signals were observed in the case of an air void. Because the void has a lower refractive index than that of the plastic, it leads to a change in the polarity of the THz pulse. The THz pulse can penetrate the surface of the plastic plate, resulting in a second pulse inside the plate. In the case of a water-filled void, most of the energy of the THz pulse is absorbed by the water. In contrast, the THz pulse is reflected by the metal defect, thus producing the largest pulse to be obtained from any of the three defects. In this case, the defect is not flat and did not disturb the entire incident THz pulse. Therefore, a reflection signal also appeared from the bottom surface. Through analysis of both the polarity and the phase information of the output signals shown in Fig. 4.5d, the types and conditions of the defects are expected to be investigated.

One of the simplest examples of THz pulsed TDI is observation of the internal structure of Japanese panel screens [3, 4]. A typical model and its cross-section as observed by THz pulsed TDI are shown in Fig. 4.6. The base paper layer is made using strong paper applied directly to the wooden lattice, and the middle preparation layer is then formed using partially glued pieces of paper. This middle preparation layer reduces the tension in the painting layer that is caused by changes

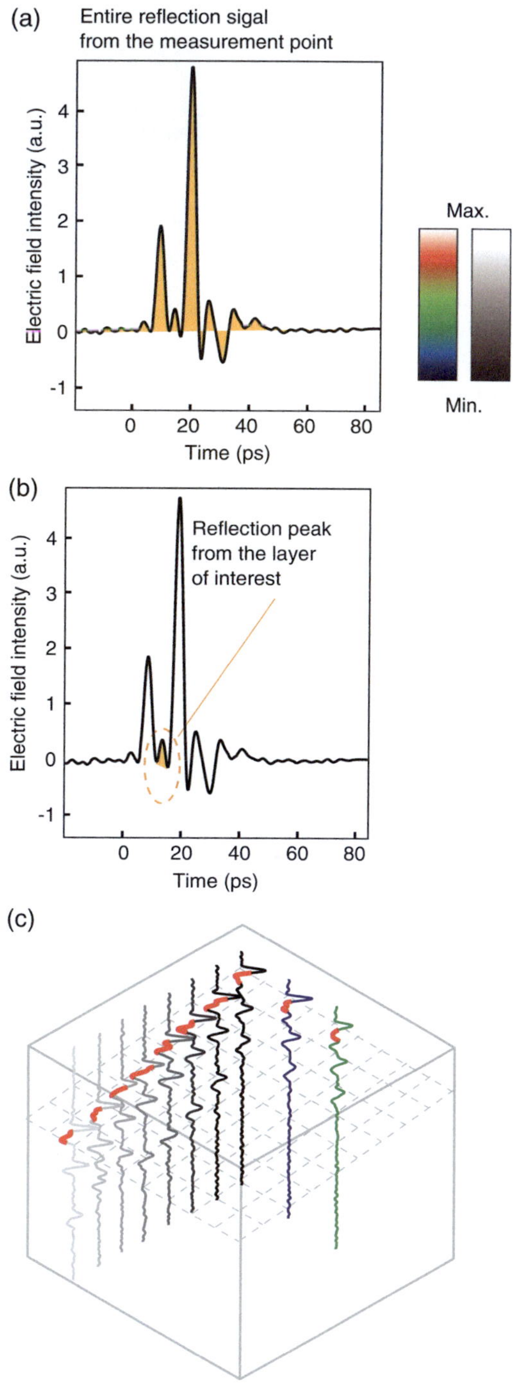

Fig. 4.3 Values used to create an area image by integrating the signal at each point. (**a**) Entire signal, (**b**) signal from the layer of interest, (**c**) acquisition of area image of the layer of interest

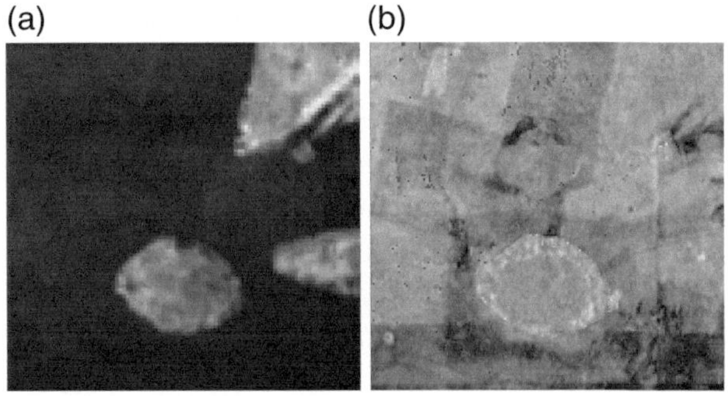

Fig. 4.4 Area image from THz reflection by power integration of part of the "Cypress Tree" by Kano Eitoku. (**a**) Entire signal, (**b**) signal from preparation paper layer. The observation process is discussed in detail in Sect. 6.5

in temperature and humidity. As shown in the figure, each piece of paper is made visible by THz pulsed TDI. Because the conservation process used for Japanese panel screens includes dismounting of the painted paper from the support, the condition of the internal structure itself is also important for conservation planning.

One practical advantage of THz pulsed TDI is the ability to detect separation in a multi-layered object. Indeed, a layer of air or metal between two dielectric materials produces the highest contrast in the imaging results. Figure 4.7 shows a painted sample on a transparent COP (Zeonex®) plate. Some parts of the sample have come away from the surface, and the separated area is only visible from the rear, as shown in Fig. 4.7b. The defect edge was filled with a type of acrylic resin. THz pulsed TDI when observed from the surface clearly revealed the existence of the air gap as a defect, as shown in Fig. 4.7c, d. The reflection from the area that is filled with the acrylic resin is higher than that from the original paint.

Separation is a major problem for Japanese paintings on paper. Figure 4.8 shows a sample with model defects that were prepared deliberately by skilled conservators. The sample has a preparation layer of gofun ($CaCO_3$ pigment from shells) on a support, similar to the gesso layer used in western panel painting. The support was made using paper glued on a wooden plate with animal glue. Separation can be generated between the preparation layer and the painting layer, and/or the support and preparation layers. The THz area image shown in Fig. 4.8b reveals internal defects as strong reflections, and the cross-section image shown in Fig. 4.8c confirms the presence of the air gap between the painting and the preparation layers.

Figure 4.9 shows a tempera painting model (Fig. 4.9a) that has been partially covered with gesso, as shown in Fig. 4.9b. Because the reflection from a metal surface is much stronger than reflections from the other dielectric materials, gold leaf under paint or gesso can be detected very easily. In this model, both the gold leaf and the gold flakes that were made by the "a missione" technique can be clearly observed. In the "a missione" technique, the image is drawn using a mixture of oil and varnish, then the gold flakes are applied. However, the spatial resolution of THz pulsed TDI imaging may not be sufficiently high to observe the image properly, as in the example shown in Fig. 4.9c. At the edge of the gesso layer, scattering reduces the reflected signal, and results in a dark line [5]. Detection of metals under

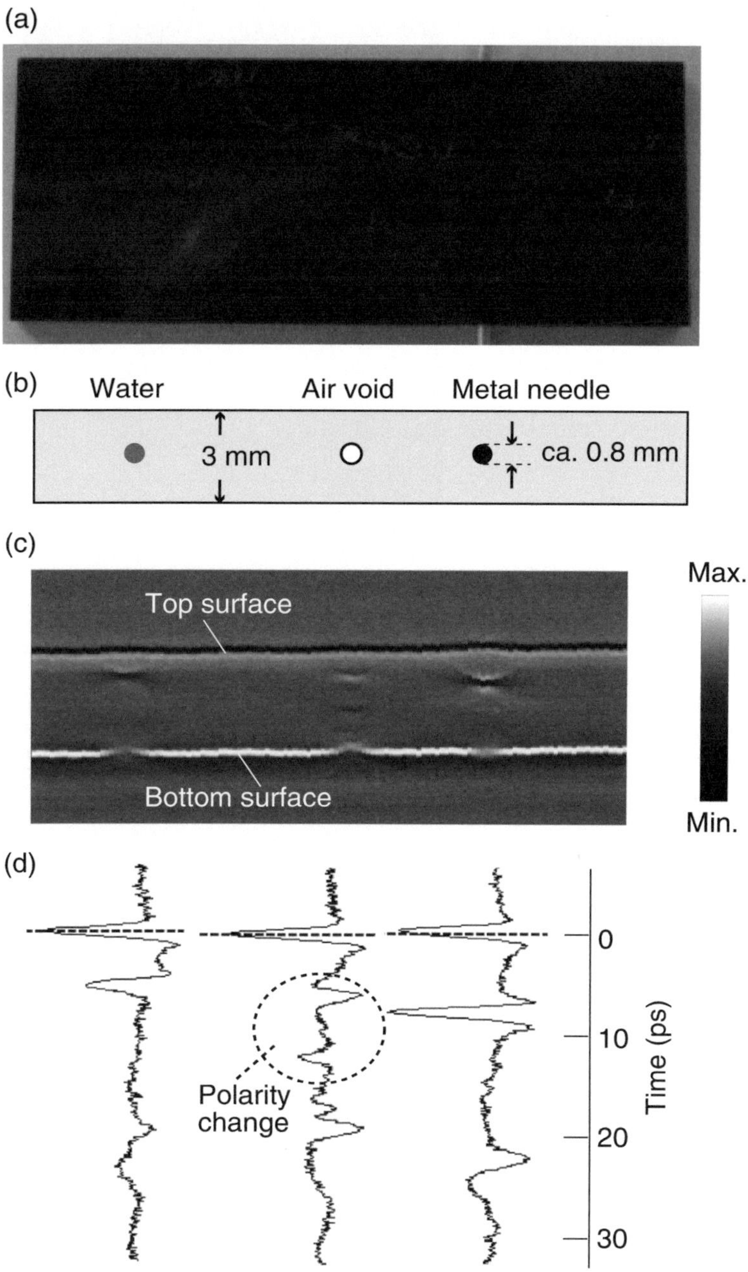

Fig. 4.5 THz reflection imaging of plastic plate sheet containing model defects. (**a**) Photograph of the model. Internal defects are not visible from the surface, (**b**) schematic depiction of defects: metal needle, air, and water, (**c**) cross-section image obtained nondestructively by THz reflection imaging, (**d**) output pulse at each point

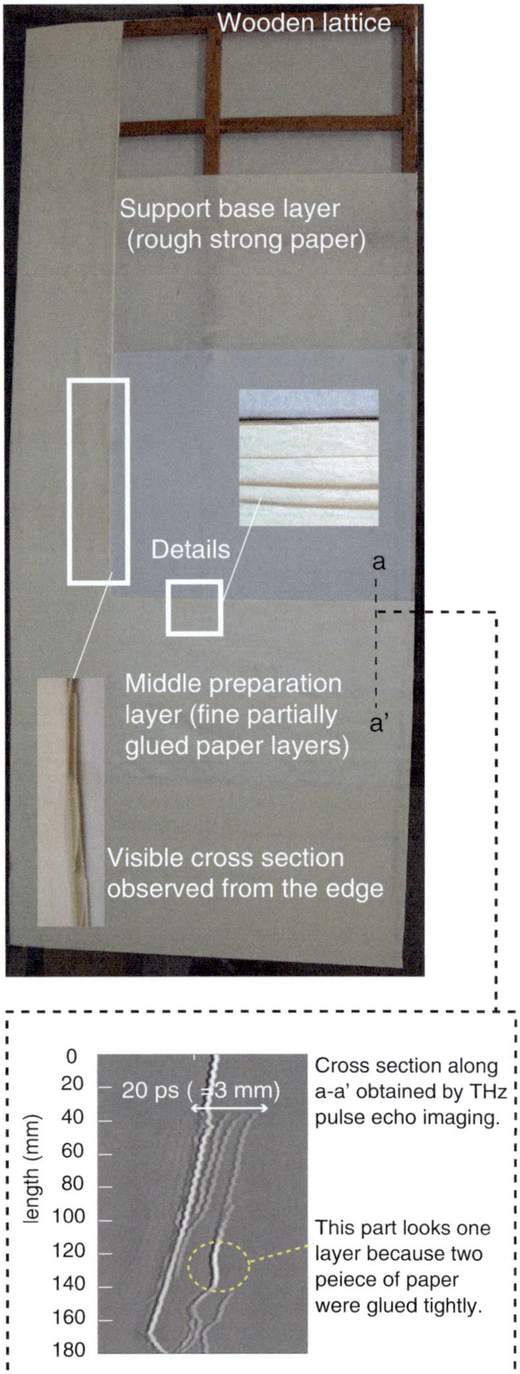

Fig. 4.6 THz reflection imaging of a Japanese panel screen model

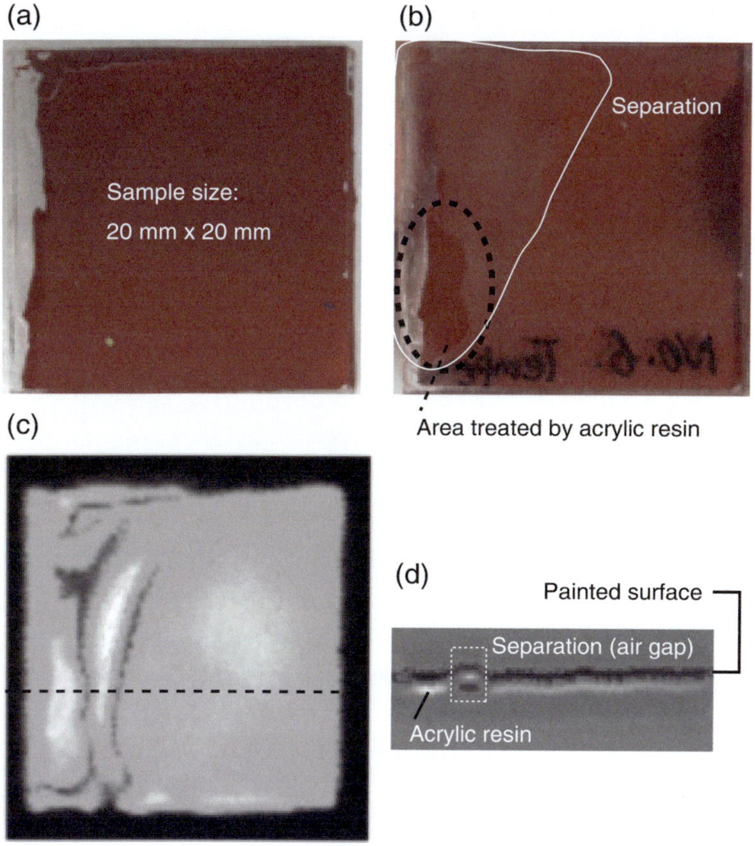

Fig. 4.7 THz pulsed TDI of painted sample on COP plate. (**a**) Sample front side, (**b**) sample viewed from rear side, with the image flipped horizontally for comparison, (**c**) THz reflection area image, and (**d**) cross-section image taken along the dotted line shown in (**c**)

dielectric materials may be regarded as too simple a task to serve as an example for experts in optical research fields, and has thus not been recognised as an important application of the THz imaging technique.

One of the most challenging objects for analysis by THz pulsed TDI is a wall painting, because the preparation layers of such paintings usually contain sand or stones of a size that is approximately the same as the wavelength of the THz wave. Figure 4.10 shows two examples of wall painting observation performed by THz pulsed TDI. The model shown in Fig. 4.10a was prepared by the buon fresco technique, which uses thick preparation layers. As shown in the THz pulsed TDI cross-section image, only the painting layer itself and the interface between the painting and the final preparation layers can be observed, and the THz pulses cannot pass through the preparation layers. In contrast, in the case of the Japanese model shown in Fig. 4.10b, the mortar preparation layer is thin enough to enable observation of the entire layer, from the surface down to the support stone. In Sect. 6.2, the THz pulsed TDI results of eight types of wall painting model are introduced.

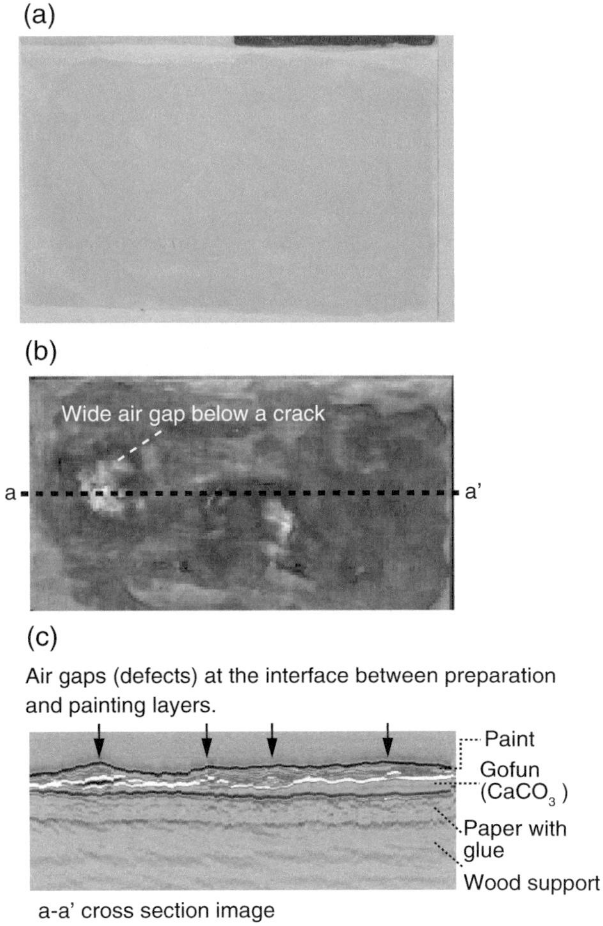

Fig. 4.8 THz pulsed TDI of painted sample on paper with model defects. (**a**) Top view of the sample, (**b**) THz area image obtained via power integration of entire waveform at each point, (**c**) cross-section image along the line a-a′, where a crack is visible from the surface

4.3 Difficulties with THz Pulsed TDI

Before case studies are shown or comparisons are made with other classical methods, the difficulties with and limitations of THz pulsed TDI are discussed in this section, using some typical examples.

Figure 4.11 shows the factors that affect THz pulsed TDI. Surface conditions, including inclination and uneven surfaces, can cause scattering, and then the incident THz wave cannot propagate into the desired object. Similar phenomena also occur inside inhomogeneous materials.

Gold is often used in artworks, and the differences between gold leaf and gold powder paint often become a focus of public attention. The THz pulse is reflected perfectly at the surface of any type of metal leaf, but the pulse is also reflected at a

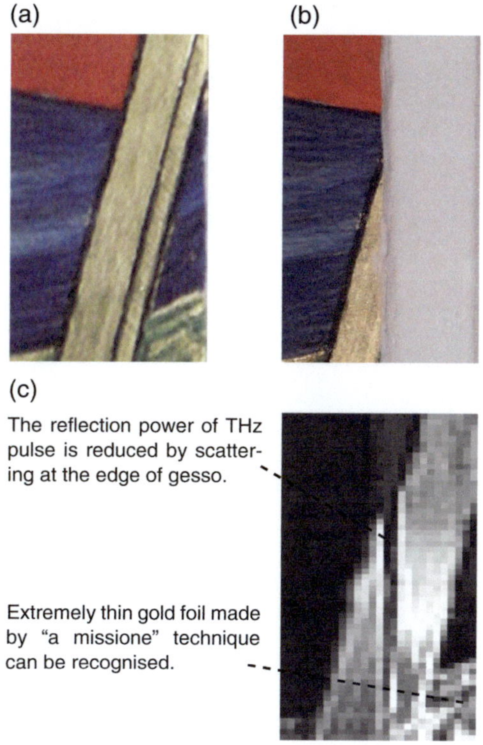

Fig. 4.9 THz pulsed TDI of a model of tempera painting that is partially covered with gesso. (**a**) Photograph before the specimen is covered with gesso, (**b**) specimen covered with gesso, (**c**) THz reflection imaging of specimen when covered with gesso

gold painted surface. As shown in Fig. 4.12, the reflection level from gold leaf is much higher than that from gold powder paint. However, when the gold leaf deteriorates, its reflection level can easily be reduced, and the leaf thus may not be easily distinguished from gold powder paint. When the oxidation of the metal, such as copper, progresses to a sufficient extent, the THz waves can pass through it, and it then appears as if it was originally made from dielectric materials. For example, transmission mode THz pulsed TDI measurements of a Sumerian bronze cup revealed that more than half of the bottom of the cup has high transmission properties. This does not mean that the highly transmitting area was made from dielectric materials. The area was originally bronze, but had already oxidised to become a dielectric material [6].

It should be noted here that the THz pulse propagation velocity depends on the refractive indexes of the materials, such that the distance from the surface then basically becomes an "apparent" value. Figure 4.13 shows an example of depth profiling of anti-corrosion paint. Figure 4.13a shows a schematic diagram of the effects of the refractive index on a cross-section image.

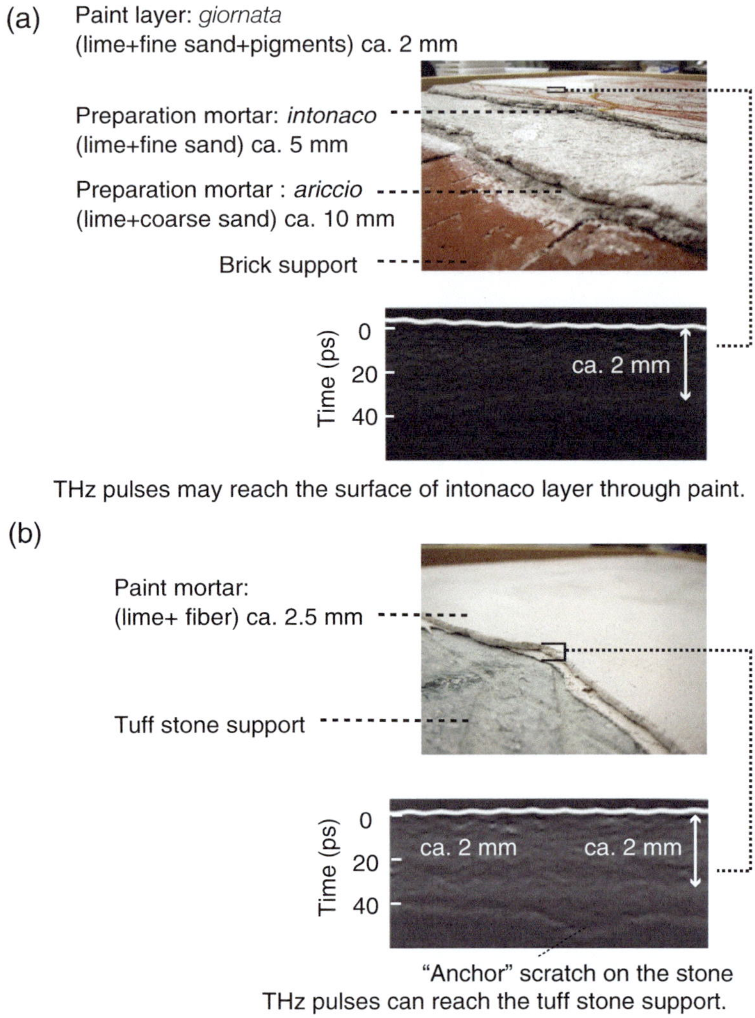

Fig. 4.10 THz pulsed TDI of a model of fresco paintings. (**a**) Buon fresco model, (**b**) Japanese model

The propagation of a THz pulse in the paint (dielectric material), for which the refractive index n_1 is higher than that of air (n_0), includes the delay from the surface, which results in the apparent distance T′ becoming larger than that required for propagation by the same distance in air, T. Because the reflected pulse from the metal surface under the paint arrives later, the metal surface in the cross-section image appears to show a bump in the metal. Figure 4.13b shows the observation area of a metal part that was painted with anti-corrosion paint where a lump was

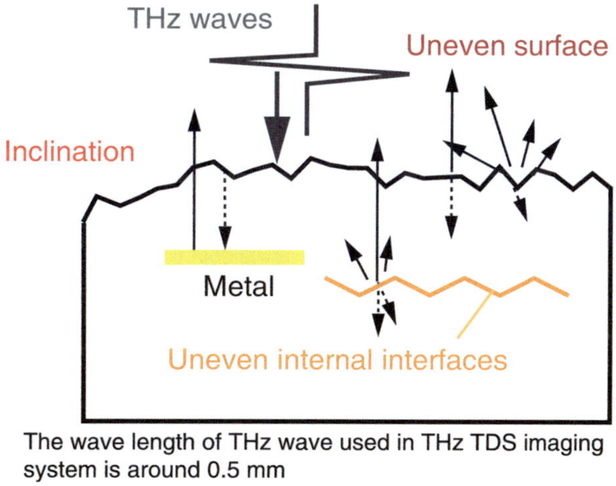

Fig. 4.11 Factors that affect THz pulse reflection imaging

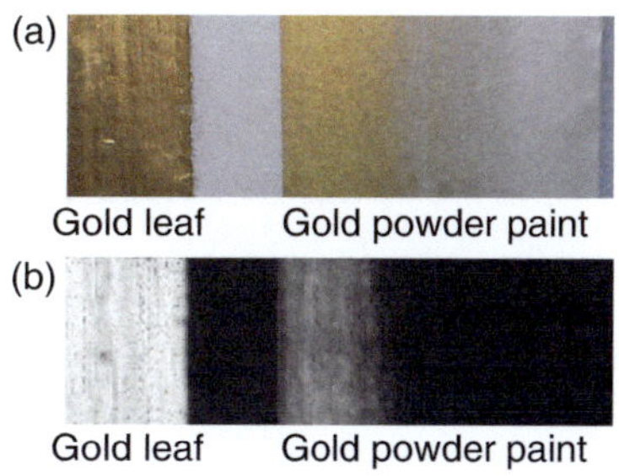

Fig. 4.12 Comparison of THz reflection imaging of gold leaf and gold paint. (**a**) Sample prepared on paper support with lime white (gofun, $CaCO_3$) preparation layer, (**b**) THz reflection image

found [7]. The THz area image and the cross-section shown in Fig. 4.13c show that the material of the lump is homogeneous, and is thus likely to be paint. If some additional layers were to appear, they would suggest the presence of rust and/or air voids. The surface condition was later confirmed by destructive testing, and no rust was found, as shown in Fig. 4.13d.

Figure 4.14 shows the difficulty of performing measurements on uneven surfaces. The object is a hanging scroll with a painting, called "Kasuga-mandara", from Nara in Japan. A THz area image that is affected by the uneven surface is shown in Fig. 4.14b. The cross-sectional image along the white broken line at least confirmed the multi-layer structure, and an additional layer on the surface was also recognised [8].

The most suitable measurement conditions for THz pulsed TDI are obtained when the angle of incidence of the THz pulse is 90°. When the object is curved, like a column, the detectable energy of the reflected pulse decreases along the line of the curvature. Figure 4.15 shows an example of observation of a painted column in Kiyomizu Temple. The scanner used was set in front of the column, as shown in

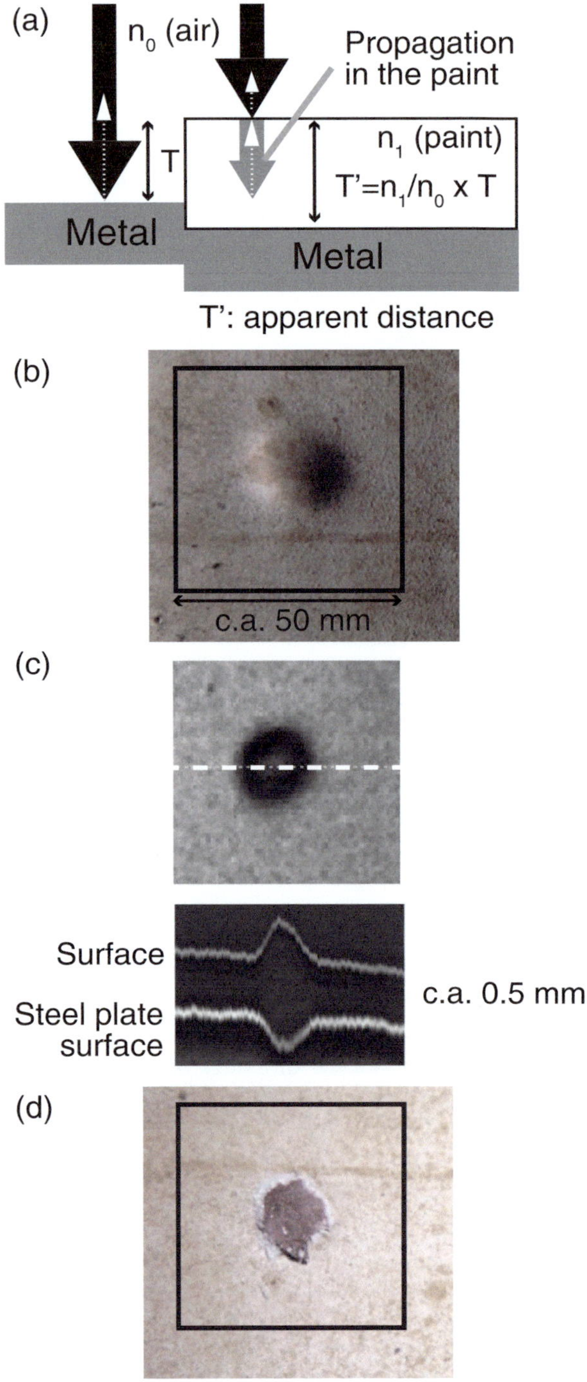

Fig. 4.13 Example of THz pulsed TDI of anti-corrosion paint on oil tank. (**a**) Schematic diagram of the delay of the THz waves, resulting in a positional shift of the same metal surface, (**b**) observation area, (**c**) THz area image and cross-section along the white broken line, (**d**) confirmation of surface condition by destructive testing

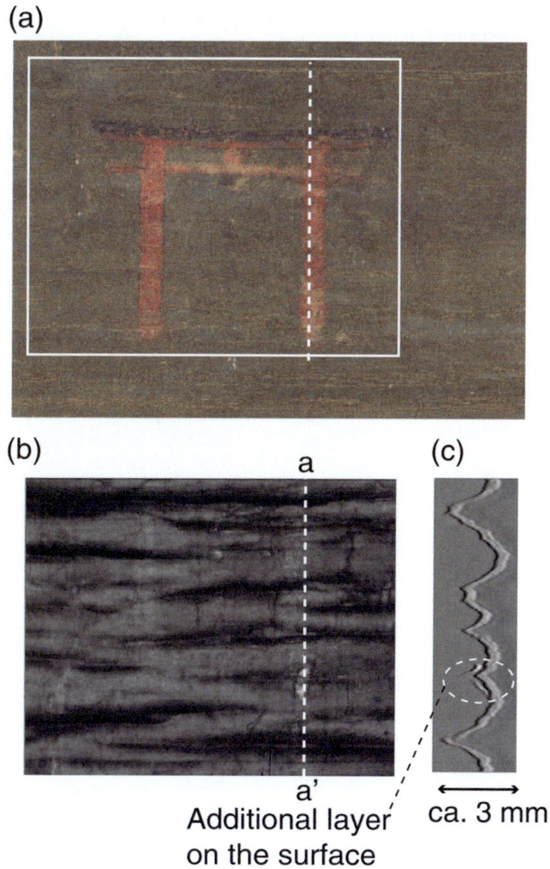

Fig. 4.14 Example of the effects of uneven surfaces on THz pulsed TDI. (**a**) Observation area of part of a hanging scroll, (**b**) THz area image affected by the uneven surface, (**c**) Cross-section image along the white broken line

Fig. 4.15a, with an observation area that is shown in Fig. 4.15b. The effect of the curvature appears clearly in Fig. 4.15c as a darker colour to indicate the reduced reflection level. However, while the signal from the edge decreases, some high reflection sections are also detected in the painted area. These sections indicate the separation of the painting layer from the wood. The experiments at Kiyomizu Temple are described in greater detail using additional data in Sect. 6.4.

If the sensor unit can be mounted on a robot or on a three-dimensional scanner, the angle of incidence of the THz pulse can be always maintained at 90° degree to the surface of the object under investigation. This does not therefore require breakthrough-type progress. Such a system will be developed when the cost matches the benefits of using THz pulsed TDI.

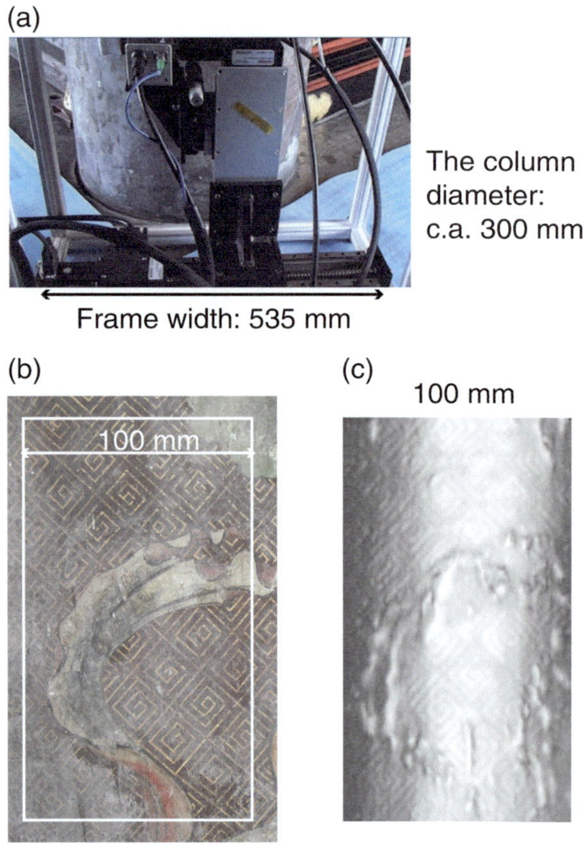

Fig. 4.15 Examples of the effects of inclination on THz pulsed TDI. (**a**) Measurement of a wooden column in a pagoda of Kiyomizu Temple, (**b**) observation area, (**c**) THz area image

4.4 Prospects for Data Post-Processing

Almost all THz experts at present have academic backgrounds in physics and optics, and thus the technology mainly progresses in terms of the development of instruments, such as high power sources. Few works have been devoted to the signal processing techniques required to extract the information from the output signals. The following examples introduce the problems and various issues related to the post-processing of the output signals for NDT.

The method used to obtain area images from THz reflection pulses at each point was shown earlier in Fig. 4.3. The power integration of the waveform or of a particular peak is often chosen as the value at each point. These values can be determined in various ways, such as the use of the maximum value or the peak-to-peak value. Figure 4.16 compares some of the processing methods used. While the difference is not too great in this case, the theoretical meaning of the image must be

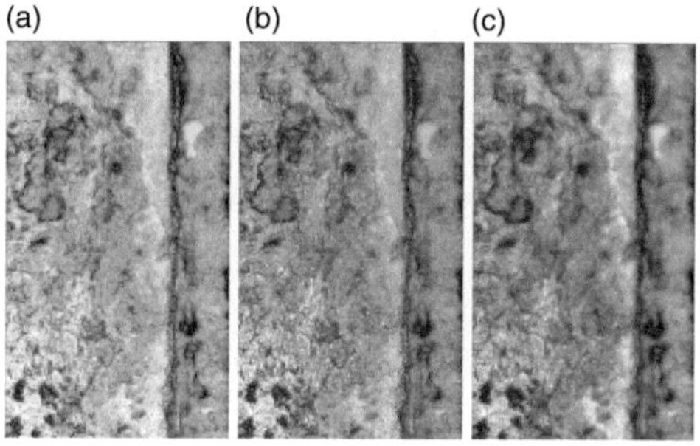

Fig. 4.16 Images obtained by different signal analysis methods. (**a**) Maximum value, (**b**) peak-to-peak, (**c**) power integration in the time domain

discussed more precisely. THz experts usually observe what they know to show the performance of their imaging systems, and choose the most appropriate measurement conditions and parameters to obtain the images that they expect. However, the users observe unknown objects, and the data must be interpreted to provide a suitable diagnosis. The meanings of the hidden values in the image must therefore be considered carefully and stated clearly.

In infrared imaging, the infrared false colour (IRFC) technique is often used and involves shifting of the longer wavelengths to the infrared region; i.e., the infrared is set to R, red is set to G and green is set to B. The IRFC technique is used in practice to distinguish between pigments that have similar colours to the naked eye, such as azurite and ultramarine, or gal ink and carbon black. At the end of this chapter, the possible use of false colour description in THz pulsed TDI is discussed. Figure 4.17 shows an example of the THz false colour model based on spectra obtained by FT-THz, and in transmission mode. Figure 4.17a shows the concept of making false colours in the THz region, with examples including the spectra of cadmium red and zinc white. The method works by dividing the applicable frequency range (2–13 THz in this case) into three parts and integrating the transmission spectra in each range. Then, the integration ratio is set to match the ratio of the RGB values to obtain the false colours. In this way, cadmium red becomes a dark green, and zinc white becomes blue. Figure 4.17b shows a model of a stained glass-like sample that was painted using various pigments. As shown in the white-painted area, the sample appears to be white to the naked eye. Using the THz false colour description method, all white pigments can be distinguished in theory [9].

If the sample's shape and physical properties are well controlled, and specific absorption peaks appear in the THz pulsed TDI frequency range, then the false colour imaging method can be useful. Pharmaceutical industries may find it useful to perform signal processing using this method to detect or observe the distributions of specific substances.

However, the operating frequency range of current THz pulsed TDI systems is limited to the range from approximately 0.1–1.2 THz, and thus practical realisation of the false colour imaging method in the near future is unlikely.

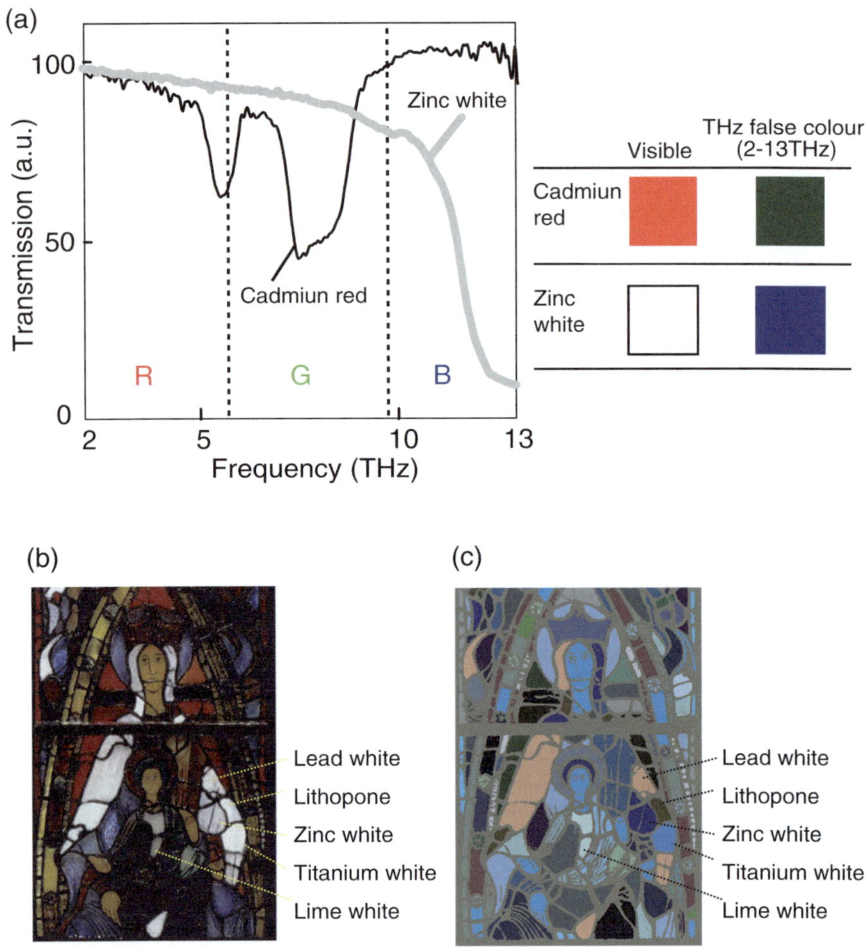

Fig. 4.17 Example of THz false colour description method using spectra obtained by the FT-THz system. (**a**) Schematic of the principle of false colour generation and examples of two pigments: cadmium red and zinc white, (**b**) model painting prepared using different pigments for similar colours, (**c**) expected false colour description of the model painting

References

1. T. Maeno, T. Futami, H. Kushibe, T. Tanaka, C.M. Cooke, IEEE Trans. Electr. Insul. **23**, 433 (1988)
2. G.C. Walker, J.W. Bowen, J.B. Jackson, J. Labaune, G. Mourou, M. Menu, W. Matthews, I. Hodder, Sub-surface terahertz imaging through uneven surfaces: visualizing neolithic wall paintings in Çatalhöyük. Proceedings of the CLEO: Science and Innovations, No. CTu3B (2012)
3. J.A. Fraser, Karibari- Japanese drying screen. The Australian Institute for the Conservation of Cultural Material. AICCM Bull. **14**, 33–50 (1988)

4. Sotheby's (Firm), *Japanese Prints, Paintings and Works of Art*, University of Minnesota Press, Mineapolis (1986)
5. G.P. Gallerano, A. Doria, E. Giovenale, G. Messina, A. Petralia, I. Spassovsky, K. Fukunaga, I. Hosako, THz-ARTE: Non-invasive terahertz diagnostics for art conservation Proceedings of the 37th IRMMW-THz, (2008)
6. J. Bianca Jackson, J. Labaune, R. Bailleul-Lesuer, L. D'Alessandro, A. Whyte, J.W. Bowen, M. Menu, G. Mourou, Terahertz pulse imaging in archaeology. Front. Optoelectron. **8**, 81–92 (2015)
7. K. Fukunaga, NDT applications of mmw-THz pulsed TDI in practice. The Japanese Society for Non-Destructive Inspection, MI symposium, (2014) in Japanese
8. M.-J. Kim Y. Kohdzuma, K. Fukunaga, Application of terahertz wave imaging technique to structural survey of a historical painting on silk, Proceedings of the Asia-Pacific microwave photonics conference, No. MC-5, (2013)
9. K. Fukunaga, Y. Ogawa, S. Hayashi, I. Hosako, Terahertz spectroscopy for art conservation. IEICE Electron. Express **4**, 258–263 (2007)

Chapter 5
Comparison of THz Pulsed TDI with Classic Methods

When introducing a new technique, it is essential to compare that technique with the existing classic methods, which have been described well in a recent review paper [1]. In this chapter, the results of THz pulsed TDI are compared with images obtained via other NDT methods that use electromagnetic waves, which operate at frequencies ranging from microwaves to X-rays. The results provide useful information about the potential and the limitations of THz technology for heritage science applications.

5.1 Nuclear Magnetic Resonance (NMR)

Depth profiling techniques using nuclear magnetic resonance (NMR) have been used in industrial applications, such as geological surveys for the oil and mining industries. NMR can be used to observe both layer thickness and porosity, based on the signals from water contained in the sample. Water is thus applied to test samples and must be maintained at a certain level during measurements. In general, NMR observations are performed using large systems and the samples are set into the system. Aachen University has developed unique single-sided NMR devices that are portable for field test applications, and these devices have been used to perform various wall painting inspections, including inspection of Pompeii-style fresco paintings. These systems are now commercially available (NMR-MOUSE, Magritek, Wellington, New Zealand).

The NMR operating frequency is the proton frequency of 13.78 MHz, which means that the penetration depth is much greater than that of THz waves, as shown in Fig. 5.1a. For comparison, the cross-section observation by THz pulsed TDI that was shown in the previous chapter in Fig. 4.1 is shown here in Fig. 5.1b. Figure 5.1c shows a schematic diagram of nondestructive depth profiling by the single-sided NMR device. When a sample has a multilayer structure, each layer has a different water saturation capacity, and this affects the intensity of the water signal that is

K. Fukunaga, *THz Technology Applied to Cultural Heritage in Practice*, Cultural Heritage Science, DOI 10.1007/978-4-431-55885-9_5

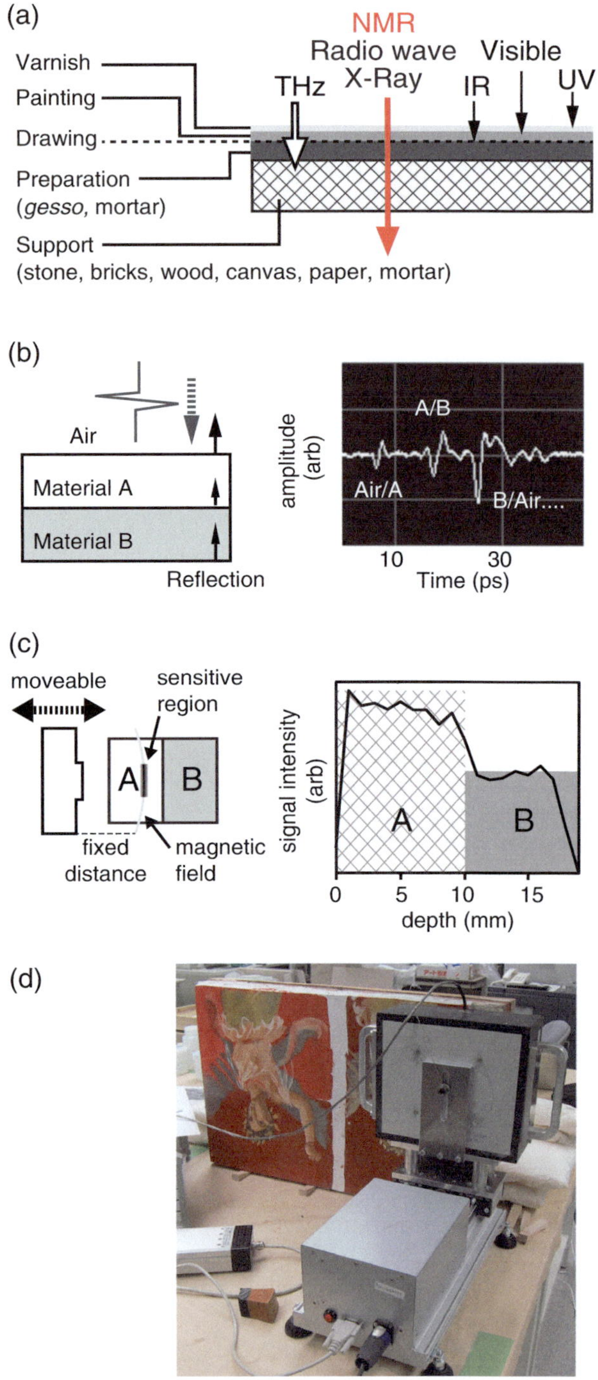

Fig. 5.1 Comparison of NMR and THz methods. (**a**) Penetration depths of electromagnetic waves, (**b**) schematic of THz pulsed TDI, (**c**) schematic of depth profiling by NMR, (**d**) photograph of portable NMR system developed by Aachen University

detected by NMR. The single-sided NMR device is only sensitive to water within a limited volume, depending on the device size. By moving the NMR device itself, the signal region can be scanned throughout the depth of the material. A detailed description of the NMR depth profiling technique can be found in the references [2–4]. Figure 5.1d shows a photograph of a wall painting measurement. Because the water that is required to increase the NMR signal actually obstructs THz signals, the THz measurements were performed first on each sample, prior to wetting of the samples for NMR.

Figure 5.2a shows a fragment of a fresco painting prepared in the traditional Pompeii style [5]. There are several preparation layers, each of which has a thickness of approximately 3 mm, as shown in the photograph in Fig. 5.2b. The two bottom layers are composed of mortar and sand, and the top preparation layer is then made using mortar and fine marble powder. The red mortar layer is formed on the marble mortar layer as a background colour, and is then polished before its surface can be painted.

The NMR system provides two output signals. One signal corresponds to the number of protons, i.e., the amount of water, and is shown as a blue line. The other signal depends on the interaction between the material and the water, and is shown as a red line. The peaks that appear in the blue line indicate the presence of interfaces where the water can accumulate. The red line profile clearly shows the material differences between the sand and marble powder mortar layers. NMR is effective for performing depth profiling of fresco wall paintings, and especially for locating internal interfaces. However, acquisition of a one-dimensional depth profile by NMR took more than 40 min. Therefore, area imaging is not recommended in practice.

In contrast, THz pulsed TDI cannot show any layer structures in the sand mortar layer, as shown in Fig. 5.2c. This is because of Mie scattering by the sand particles, which are of similar size to the wavelengths of the electromagnetic waves in the THz range. Also, the painting layer is too thin to be distinguished from the preparation layer by THz pulses. Additional examples of THz pulsed TDI for wall paintings are given in Sect. 6.2.

5.2 Microwave and Millimetre-Wave Radar Imaging

Below the THz frequency region, radar technology using microwaves is well established and has been used in archaeological field research. Major examples include ground penetrating radar for the detection of buried objects, and the various aerial and satellite remote-sensing technologies used to discover and search ruins around the world [6]. Microwave radar technology has been applied in infrastructure observation, and millimetre-wave technology is under development for use in the observation of targets located relatively near the surface, such as corrosion under paint [7–9]. For conservation planning of wall paintings, the condition of the supporting wall itself is also an important factor, and microwaves and millimetre-

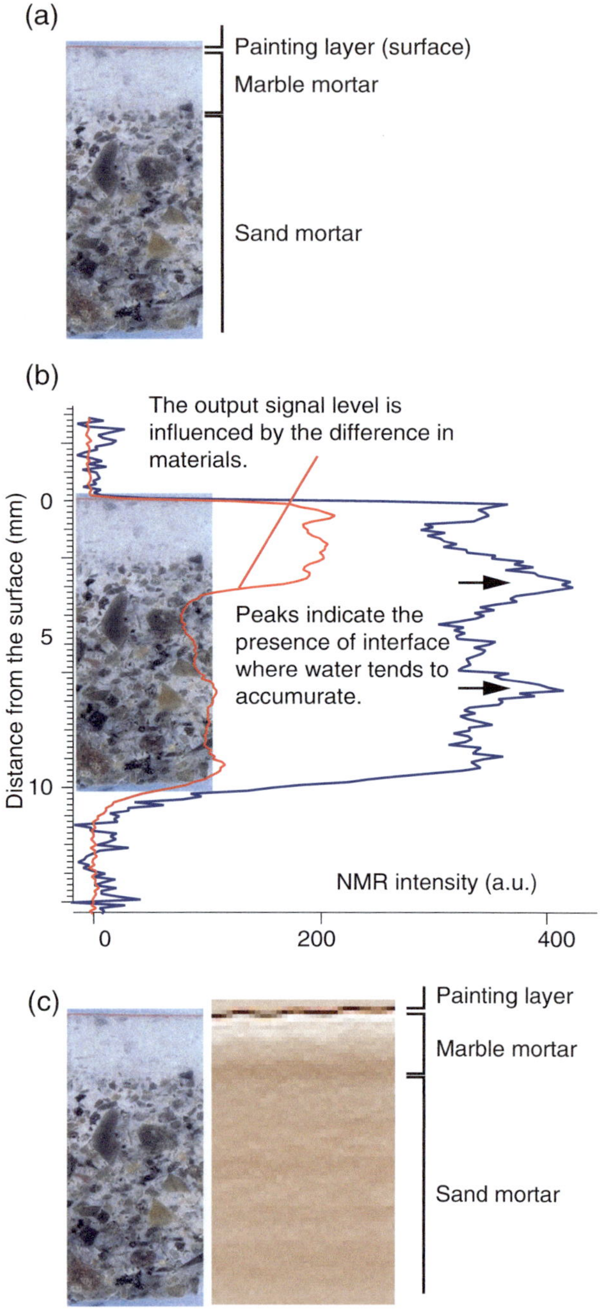

Fig. 5.2 Comparison of NMR depth profile with THz cross-section imaging. (**a**) Microscopic observation of fragment of Pompeii-style fresco painting model, (**b**) NMR depth profile in two modes, where the blue line is the water-sensitive mode, and the red line corresponds to material properties, (**c**) THz cross-section image

waves are expected to be used to observe the internal structures of walls. Figure 5.3 shows a comparison of damaged wall inspections performed by microwave, millimetre-wave, and THz pulsed TDI methods. The observation area shown in Fig. 5.3a has a narrow visible hairline crack. The microwave radar used in this work is a multi-pass linear array radar (MPLA245A, Mitsui Engineering & Shipbuilding, Tokyo, Japan) that has been commonly used in the civil engineering field to inspect roads and bridges. The entire wall area (1,000 × 1,500 mm) is measured in 5 min, with measurement resolution of approximately 30 mm. The observation area by millimetre-wave is indicated by the yellow squares shown in Fig. 5.3b. The wavelength of the microwaves is too long to enable observation of every surface of the wall, e.g., at depths of less than 20 mm, while rough profiles can be observed to depths of up to 300 mm. The blue-green coloured area appearing in Fig. 5.3b is a high reflection area that indicates the presence of iron pipes or an air gap. A three-dimensional image of the area indicated by the black broken squares in Fig. 5.3b is shown in Fig. 5.3c. There is a diagonal highly reflective line along the hairline crack that is not visible by microwave radar imaging. This indicates that the hairline crack was generated along the iron pipe.

Figure 5.3d, in contrast, shows the image obtained using the millimetre-wave imaging system developed by SynView Ltd., which was described earlier in Fig. 2.6. The imaging time for a 400 mm square is 10 min with spatial resolution of 3 mm. This system is shown in Fig. 2.6. High reflection levels around the hairline crack indicate the presence of an air gap, which may cause the tiles to come off the surface. In the case of THz pulsed TDI, as shown in Fig. 5.3e, a separation of the tiles near the hairline crack is observed with resolution as high as 1 mm, although it does take around 30 min to produce this tiny image. It is thus not very practical to apply THz pulsed TDI for entire wall inspections.

This comparison has proved that microwaves are useful for investigating the internal structure of the wall itself. In historical monuments, many walls have been covered by new layers for various reasons, including changes in the religious body of the monument. One of the most famous of these "covered walls" is the internal wall of the Hagia Sophia in Istanbul. The walls were originally made with precious mosaic art coverings in the Byzantine era, and were later covered with gesso plaster. While the presence of the mosaics was discovered during conservation works in 1848, the art was covered again until the 1930s. Some parts of the mosaics are now visible after the removal of the gesso layer by the Byzantine Institute [10].

Because microwaves can pass through approximately 10-mm-thick gesso layers, the instrument that obtained the image shown in Fig. 5.3b was applied to observation of some parts of the Hagia Sophia during a research project led by Prof. K. Hidaka of the University of Tsukuba. Figure 5.4a shows the microwave reflection image of one part of the Hagia Sofia where a mosaic geometrical pattern (shown in Fig. 5.4b) was found nearby. The high reflection parts of the image suggest the presence of gold or silver tesserae, some metallic parts used in previous conservation efforts, or air voids. Based on the distribution of tesserae nearby, as shown in Fig. 5.4b, the square and round patterns that appeared in Fig. 5.4a are considered to be part of successive geometrical patterns [11].

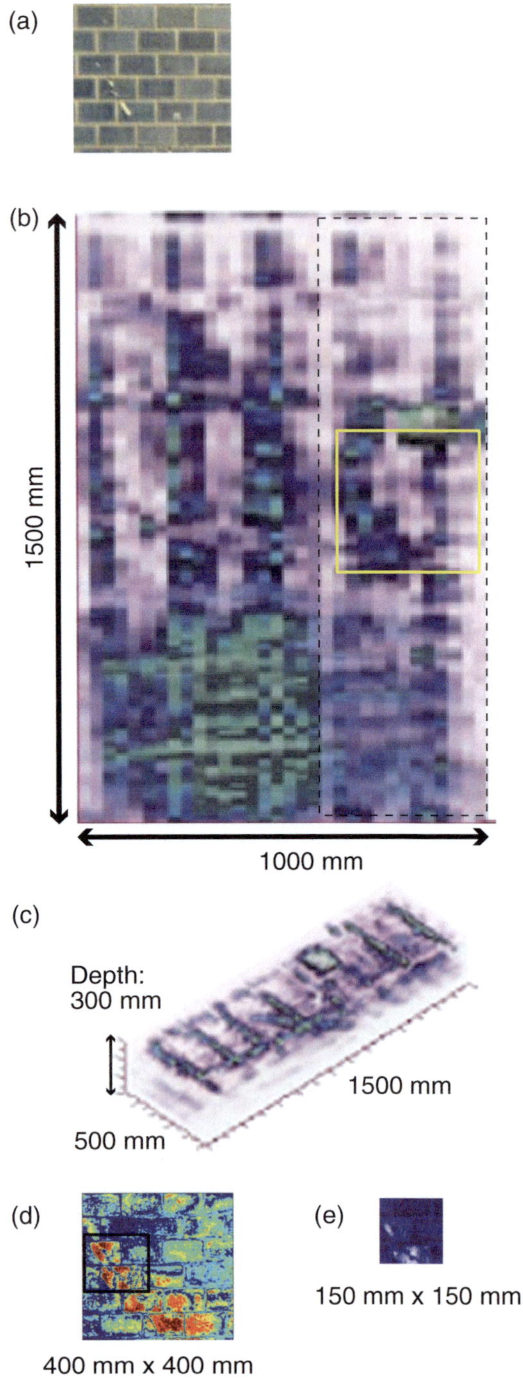

Fig. 5.3 Comparison of wall inspections performed using electromagnetic waves. (**a**) Observation area with hairline crack, (**b**) microwave radar observation at 5 GHz, (**c**) three-dimensional image of the area indicated by black broken lines in part (**b**, **d**). Millimetre-wave continuous wave imaging at 100 GHz in the area indicated by the yellow square in part (**b**, **e**). THz time domain observation of the area indicated as a black solid square in part (**d**)

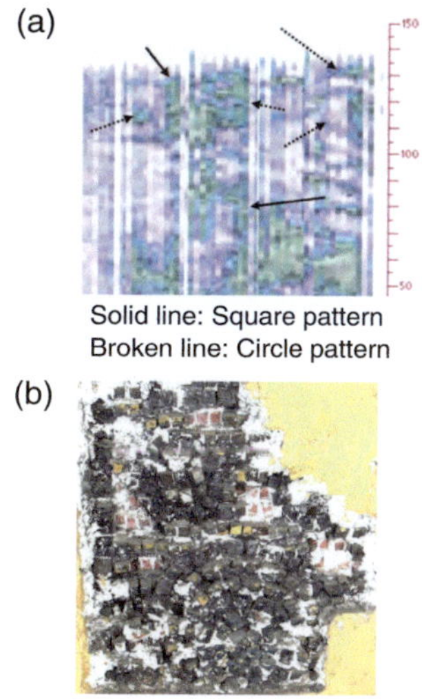

Fig. 5.4 Observation of part of a wall in the Hagia Sofia. (**a**) Microwave reflection image at 4 GHz, (**b**) mosaic pattern under gesso plaster next to the observation area shown in part (**a**)

5.3 Mid- and Near Infrared (NIR)

Above the THz region, the infrared region has been commonly used by infrared cameras, and NIR cameras have widely been used for inspecting agricultural products. Multispectral imaging generally refers to an imaging system with frequency coverage from the mid-infrared to UV. In that range, Raman spectroscopy has also been actively used in heritage science [12–15].

Figures 5.5 and 5.6 compare observations of paintings on wood by THz pulsed TDI and by NIR imaging with a selectable light source. The photograph shown in Fig. 5.5a is a painted internal wall in the pagoda of Aizendo (built in 1597), which is the oldest wooden structure in Osaka. In Japanese paintings, the lines are basically drawn using carbon black, and as a result, are invisible to THz pulsed TDI, as shown in Fig. 5.5b. However, the pale yellow sharp lines are much clearer than those shown in the visible photographs. This proves that the lines are composed of gold leaf. While the paint layer is too thin to be recognised as an independent layer on the wood, an air gap was clearly observed between the paint and the wood, because the reflection at the interface between paint and air is much higher than that between the paint and the wood. The cross-section image shown in Fig. 5.5c also indicates the separation of the painting layer from the wood support. However, when the paint came off completely, the area appeared dark because of the low reflection from the wood surface.

Figure 5.6 shows NIR images of an area close to the region where the THz pulsed TDI process was performed. The same area was observed at wavelengths of 900, 1200, and 1550 nm. The best contrast is obtained at the wavelength of 1,200 nm, and in all cases, the carbon black lines appeared clearly, while the gold leaf cannot be distinguished from other painted areas.

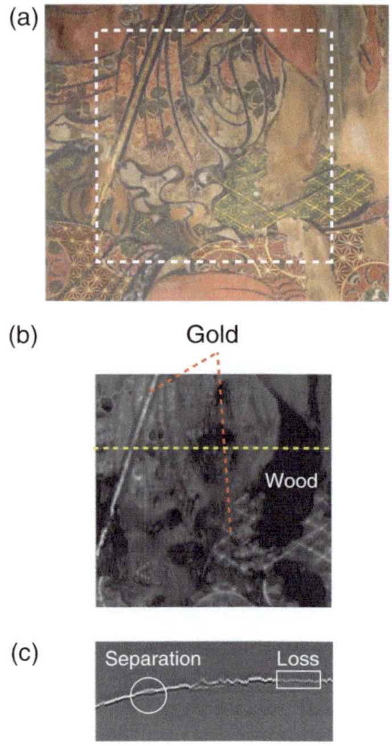

Fig. 5.5 THz pulsed TDI of a painting on wood. (**a**) Observation area, (**b**) THz reflection area image, (**c**) cross-section image along the yellow dotted line in (**b**)

In addition to these classical uses of infrared light, optical coherence tomography (OCT) technology has also progressed rapidly. Recent results indicate the high potential for application of OCT to artwork analysis [16–19].

5.4 Visible Microscopy

The most fundamental and direct observations occur in the visible region. If a tiny sample is allowed to be taken, microscope observations of the cross-section of the sample give almost all required information about painting techniques and conditions, although the number of sampling points is obviously limited. Figure 5.7a, b show macro- and microscopic observations of a fragment of a fresco painting prepared in the Pompeii style, respectively. The cross-section image that was obtained by THz pulsed TDI shown in Fig. 5.7c corresponds to the interface between the two mortar layers: one contains marble powder, while the other contains sand as aggregate, as shown in Fig. 5.7b. Additionally, an interface appears inside the marble mortar layer. This is not visible to the naked eye. This internal interface was generated during the making of this layer by painting twice. THz pulsed TDI may have detected the difference in density in this case. According to a fresco artist, there could also be a difference in the density at the interface if the first turn is made at relatively high pressure.

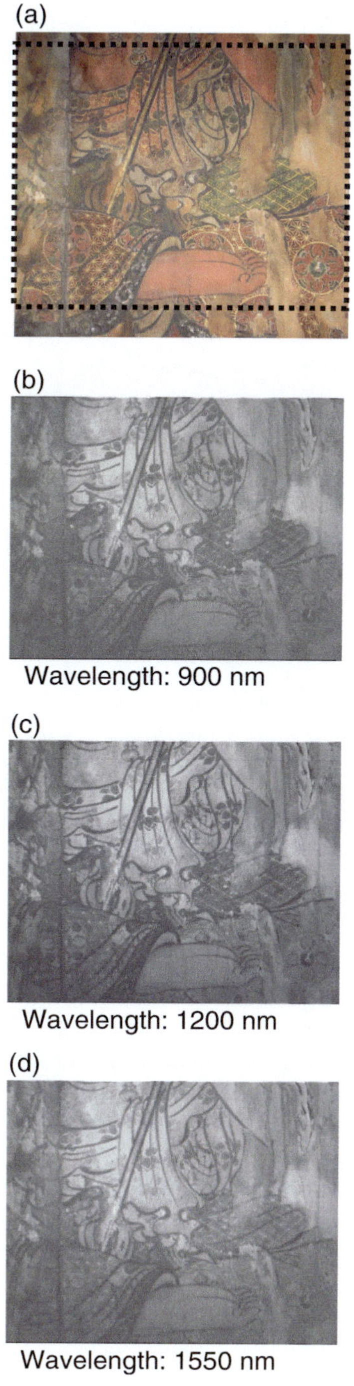

Fig. 5.6 NIR imaging of a painting on wood. (**a**) Observation area, (**b**, **c**, **d**) NIR reflection images at wavelengths of 900, 1200, and 1550 nm, respectively

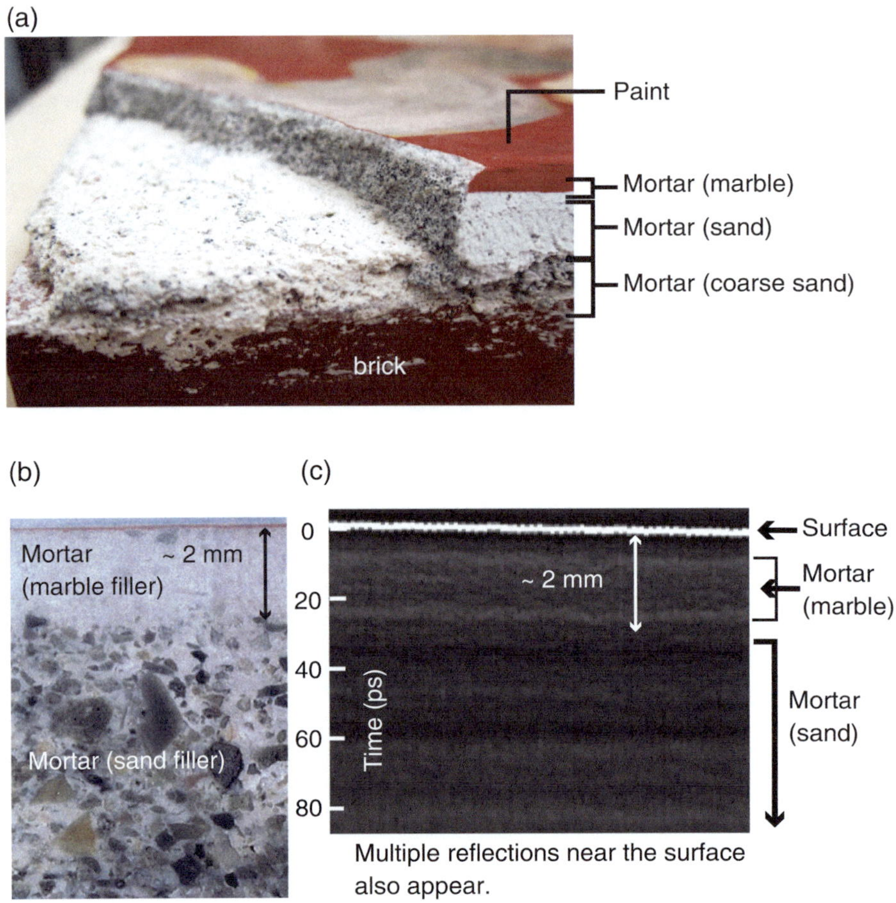

Fig. 5.7 Comparison of THz pulsed TDI with visible macroscopic and microscopic observations. (**a**) Specimen prepared by traditional Pompeii style fresco painting, (**b**) microscopic cross-section analysis of a sample in the visible range, (**c**) THz cross-section image

Figure 5.8 shows a microscopic observation of the cross-section of "Polittico di Badia", which is a masterpiece of tempera panel painting by Giotto. This sampling observation (by the micro-destructive method) was performed to confirm the presence of two gesso layers that were indicated by THz pulsed TDI. Two gesso layers are visible, as shown in Fig. 5.8b. The complete investigation of this masterpiece can be found in the literature [20], and the study is briefly described in Sect. 6.3.

In addition to microscopic observations, transmission images using visible light are useful in paper-based artworks, such as East Asian screens. Figure 5.9 shows part of the Japanese panel screen "Cypress Tree", which is a masterpiece by Kano Eitoku that is housed in the collection of the Tokyo National Museum. The THz

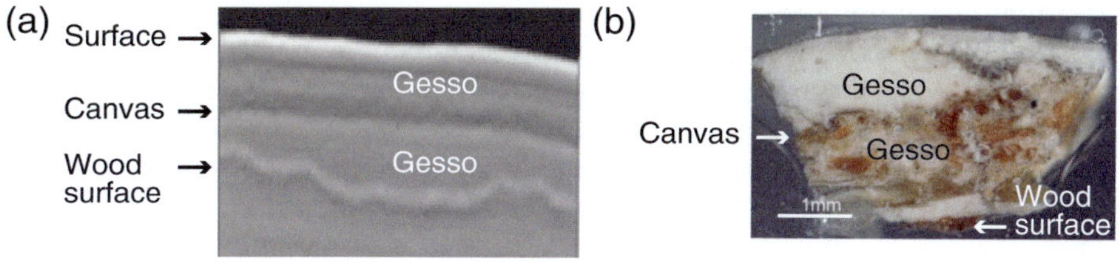

Fig. 5.8 Comparison of THz pulsed TDI with visible microscopic observation of "Polittico di Badia". (**a**) Nondestructive cross-section obtained by THz pulsed TDI, (**b**) microscopic cross-section analysis of a sample, observed by the Istituto per la Conservazione e la Valorizzazione den Beni Culturali

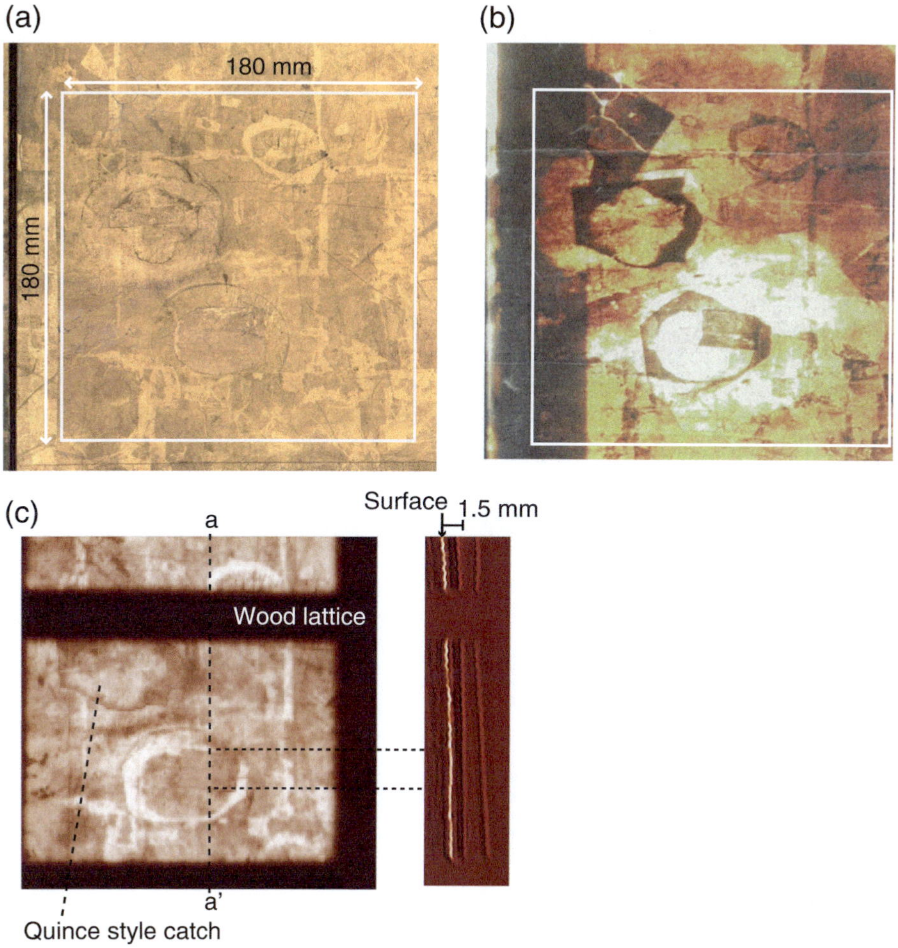

Fig. 5.9 Comparison of THz pulsed TDI with visible transmission images. (**a**) Observation area of part of a Japanese panel screen, "Cypress Tree", by Kano Eitoku from 1680, (**b**) visible light transmission image taken by the Tokyo National Museum, (**c**) THz area image by power integration from the rear side

reflection image observed from the rear side revealed various types of mending papers at the position of the former catch, as shown in Fig. 5.9b. The wood lattice mounting interrupts observation of the front side of the screen. The images observed by light transmission (Fig. 5.9c) can show most of the mending papers. While THz pulsed TDI can add information about the positions of the mending papers in the layer structure, the transmission image is often sufficient for conservation planning in artworks made using paper and glue. Details of this work are described in Sect. 6.5.

5.5 Ultraviolet Imaging

UV imaging is also widely used to observe the condition of varnishes and materials that emit fluorescence. Because the observation area for UV imaging is limited to the surface, comparisons between UV and THz pulsed TDI are omitted here.

5.6 X-Ray Radiography

X-ray radiography is widely used as an established examination method for art object analysis in many large museums around the world. X-ray CT is also commonly used to observe the internal structures of three-dimensional objects, such as statues. Recent papers have drawn clear comparisons between X-ray CT and THz CW tomography [21, 22]. Figure 5.10 shows a comparison between THz pulsed TDI and X-ray CT using a wrapped mummy as an example. X-ray CT

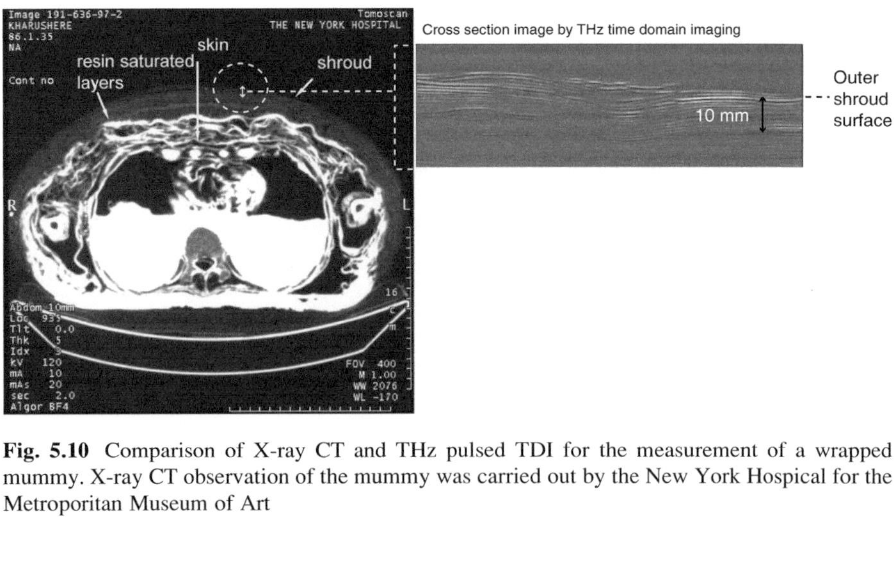

Fig. 5.10 Comparison of X-ray CT and THz pulsed TDI for the measurement of a wrapped mummy. X-ray CT observation of the mummy was carried out by the New York Hospical for the Metroporitan Museum of Art

clearly shows the interior of the body of the mummy. The wrapping clothes are transparent under conditions that have been adjusted to observe the body itself, while THz pulsed TDI can see details of the wrapping that lie very close to the surface. The THz waves only penetrate the shroud, and are absorbed in the resin-saturated layers. Details of the work performed on the mummy are discussed in Sect. 6.1. These results indicate that the spatial resolution of THz pulsed TDI cannot reach the levels of X-ray radiography or tomography, but show that the THz waves can detect organic materials that are relatively difficult to observe using X-ray techniques, because of the high transmission feature of X-rays.

5.7 X-Ray Materials Analysis

While the principal role of X-rays in heritage science in the early days was in radiography for imaging applications, X-ray element analysis is also very effective for materials identification and is now commonly used. Element analysis at a specific point is performed using a handy XRF system (e.g., Niton XL-3t, Thermo Fisher Scientific, Waltham, MA, USA). The element information is obtained from an area of less than 1 mm^2, and thus XRF-based imaging can take hours to obtain the required element map. Figure 5.11 shows a comparison between the THz pulsed TDI and XRF element mappings of a fragment of a sixteenth century mural painting from the Dazhao Monastery in Inner Mongolia. The reflection image showed that the orange colours used in the belt and skirt in the figure were painted using different materials, because the reflection levels of these two parts were different for each area. In contrast, the red and yellow colours show little difference in the THz reflection images. While THz pulsed TDI can show the differences between materials in some cases, as shown in this example, the identification of unknown materials is extremely difficult because of the lack of a spectral library and insufficient theoretical peak assignment research.

When the observation times are compared, however, THz pulsed TDI of the entire fragment, measuring approximately 150×150 mm^2, takes 30 min or less, while the XRF element mapping takes more than 5 h. A quick scan by THz pulsed TDI can thus indicate the points that should be measured by XRF or other precise analytical methods. This work is described in greater detail in Sect. 6.2 [23].

Another materials analysis method that uses X-rays is energy-dispersive X-ray analysis (EDX), which is also known as electron probe micro analysis (EPMA). While this method requires micro-sampling, the element mapping of the cross-section is very effective for understanding of the materials and the techniques used in the artwork. Figure 5.12 shows a comparison of the THz pulsed TDI cross-section and EPMA analysis of a fragment of a Pompeii-style fresco painting, which is the same sample used for the experiments shown in Figs. 5.2 and 5.6. The EDX analysis was performed in areas A and B indicated in the photograph shown in Fig. 5.12a. It is obvious that the THz reflection image cannot distinguish between the painting layer and the red preparation layer in area A. When using EDX, in

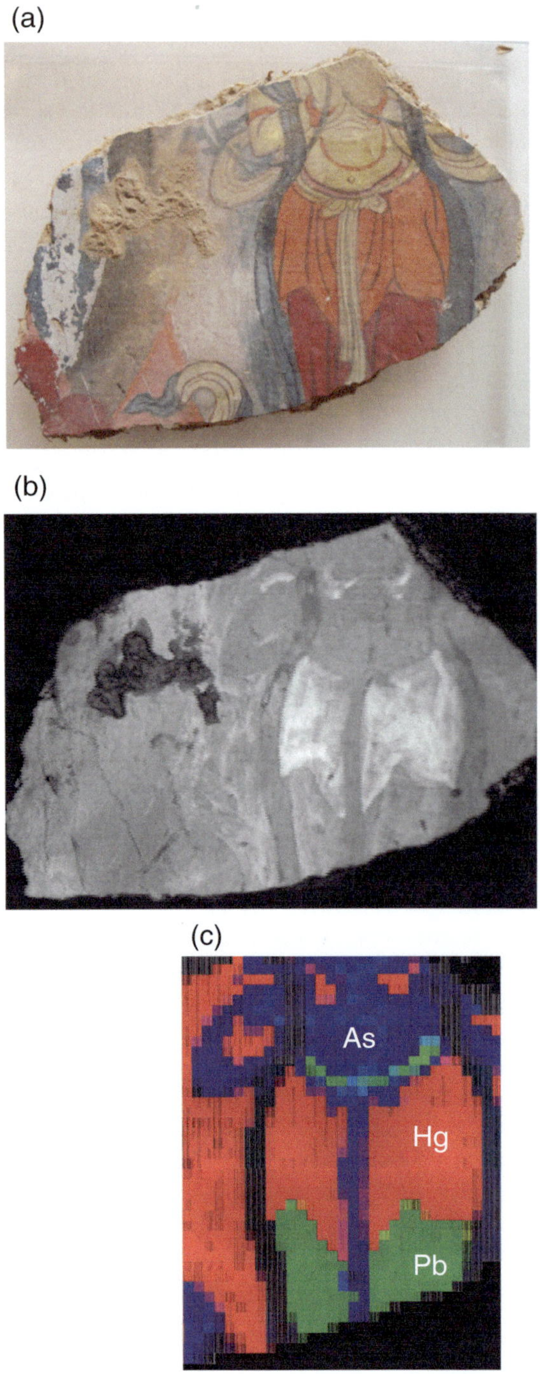

Fig. 5.11 Comparison of THz pulsed TDI with XRF element mapping of a fragment of a mural painting from the Dazhao Monastery in Inner Mongolia from the sixteenth century. (**a**) Photograph of a fragment, (**b**) THz reflection image, (**c**) XRF element mapping, taken by the Nara National Research Institute for Cultural Properties

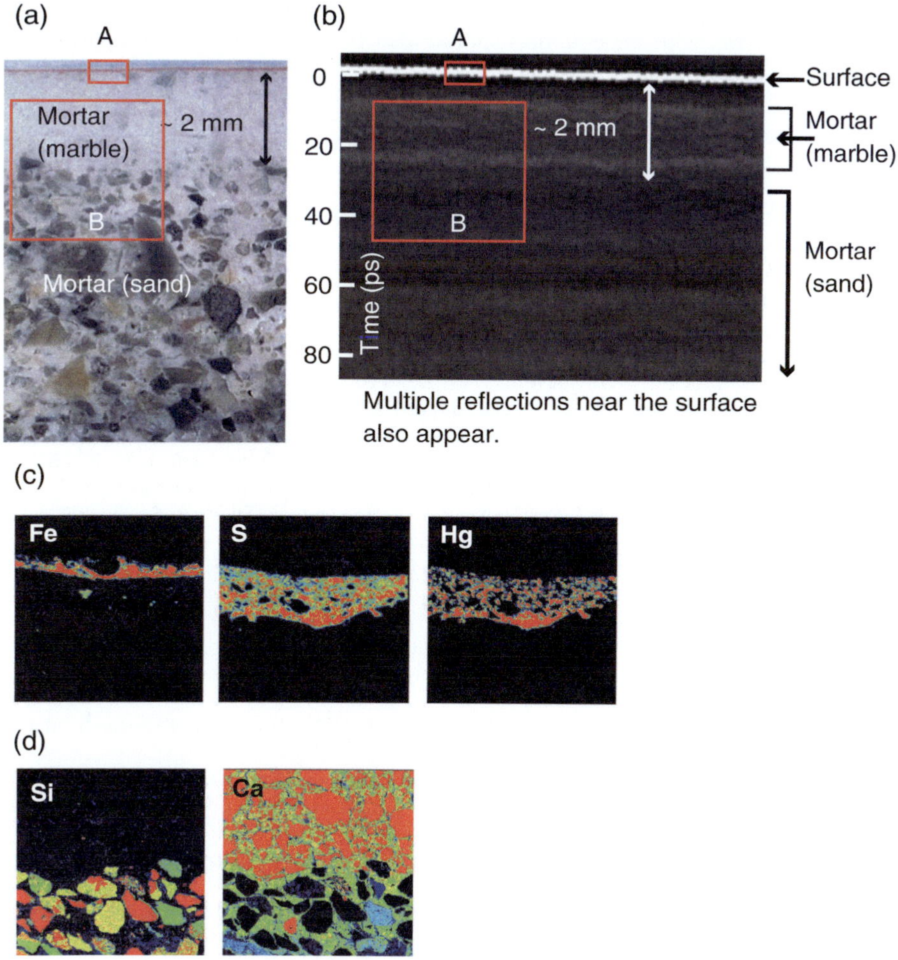

Fig. 5.12 Comparison of THz pulsed TDI with visible microscopic observation and element analysis mapping using an electron probe micro analyser (EPMA), which is also known as EDX. (**a**) Microscopic cross-section observation of a sample, (**b**) THz cross-section image, (**c**) elements mapping of area A, (**d**) elements mapping of area B

contrast, the element distribution can clearly be observed in both areas A and B, as shown in Fig. 5.12c, d. In area A, the iron on the surface is from the yellow ochre in the painting, while the mercury and sulphide come from cinnabar (mercury sulphide), which is used for the shiny red preparation. In area B, EDA analysis shows the differences between the mortar layers, which correspond to the THz cross-section (Fig. 5.12b). The difference in the materials between mortar with marble and mortar with sand is also clearly observed by THz pulsed TDI.

In addition to the electromagnetic wave-based techniques mentioned above, acoustic methods with air coupling should also be considered for comparison in the near future [24, 25].

References

1. S. Legrand, F. Vanmeert, G Van der Snickt, M. Alfeld, W. De Nolf, J. Dik, K. Janssens, Herit. Sci. **2**(13) (2014)
2. B. Blümich, F. Casanova, J. Perlo, F. Preciutti, C. Anselmi, B. Doherty, Noninvasive testing of art and cultural heritage by mobile NMR. Acc. Chem. Res. **43**, 761–770 (2010)
3. A. Haber, B. Bluemich, D. Souvorova, E. Del Federico, Ancient roman walls mapped non-destructively by portable NMR. Anal. Bioanal. Chem. **401**, 1441–1452 (2011)
4. K. Fukunaga, T. Meldrum, W. Zia, M. Ohno, T. Fuchida, B. Blümich, Nondestructive investigation of the internal structure of fresco paintings. Proc. Digit. Herit. **1**, 81–88 (2013)
5. M. Vitruvius Pollio, De Architecture (29–23 BCE), ed. L. Cherubini, P. Giardini (1975)
6. J.R. Wiseman, F. El-Baz (eds.), *Remote Sensing in Archaeology* (Springer, Berlin, 2007)
7. H. Togo, T. Kojima, S. Mochizuki, N. Kukutsu, Millimeter-wave imaging for detecting surface cracks on concrete pole covered with bill-posting prevention sheet. NTT Tech. Rev. **10**, 1–6 (2012)
8. G.G. Diamond et al., Remote detection of corrosion under paint (CUP) from distances greater than 5 metres, Proceedings of the 11th ECNDT (2014)
9. FP7 Project, Innovative non-destructive corrosion under paint integrated detection system, http://www.cupidndt.com/
10. N.B. Teteriatnikov, *Mosaics of Hagia Sophia, Istanbul* (Dumbarton Oaks Research Library and Collection, Washington, DC, 1998)
11. H. Takanezawa, K. Hidaka, Consideration of the plaster-covered mosaic on the south wall of east bay in the South Gallery of Hagia Sophia in Istanbul. Proceedings of the Architectual Institute of Japan Annual Convention, (2014), pp. 797–798
12. L. Cséfalvayová, M. Strlič, H. Karjalaine, Quantitative NIR chemical imaging in heritage science. Anal. Chem. **83**, 5101–5106 (2011)
13. R. Hedjam, M. Cheriet, Historical document image restoration using multi spectral imaging system. Pattern Recog. **46**, 2297–2312 (2013)
14. R.J.H. Clark, The scientific investigation of artwork and archaeological artefacts: Raman microscopy as a structural, analytical and forensic tool. Appl. Phys. A **89**, 833–840 (2007)
15. J.M. Madariaga, D. Bersani (eds.), Special issue: Raman spectroscopy in art and archaeology. J. Raman Spectrosc. **43**, 1523–1844 (2012)
16. P. Targowski, B. Rouba, M. Wojtkowski, A. Kowalczyk, The application of optical coherence tomography to non-destructive examination of museum objects. Stud. Conserv. **49**, 107–114 (2004)
17. D.C. Adler, J. Stenger, I. Gorczynska, H. Lie, T. Hensick, R. Spronk, S. Wolohojian, N. Khandekar, J.Y. Jiang, S. Barry, A.E. Cable, R. Huber, J.G. Fujimoto, Comparison of three-dimensional optical coherence tomography and high resolution photography for art conservation studies. Opt. Express **15**, 15972–15986 (2007)
18. C.S. Cheung, M. Spring, H. Liang, Ultra-high resolution Fourier domain optical coherence tomography for old master paintings. Opt. Express **23**(8), 10145–10157 (2015)
19. FP7 project: OCT4ART (optical coherence tomography for examination of works of art, www.oct4art.eu
20. A. Tartuferi (ed.), *Giotto, Il Restauro del Polittico di Badia* (Mandoragora, Firenze, 2012)

21. E. Abraham, M. Bessou, A. Ziéglé, M.-C. Hervé, L. Szentmiklósi, Z. Kasztovszky, Z. Kis, M. Menu, Terahertz, X-ray and neutron computed tomography of an eighteenth dynasty Egyptian sealed pottery. Appl. Phys. A **117**, 963–972 (2014)
22. J.B. Jackson, J. Labaune, R. Bailleul-Lesuer, L. D'Alessandro, A. Whyte, J.W. Bowen, M. Menu, G. Mourou, Terahertz pulse imaging in archaeology. Front. Optoelectron. **8**(1), 81–92 (2015)
23. K. Fukunaga, I. Hosako, Y. Kohdzuma, T. Koezuka, M.-J. Kim, T. Ikari, X. Du, Terahertz analysis of an East Asian historical mural painting, J. Eur. Opt. Soc. Rapid. Publ. **5**(10024) (2010)
24. A.M. Siddiolo, L. D'Acquisto, A.R. Maeva, R.G. Maev, Wooden panel paintings investigation: an air-coupled ultrasonic imaging approach. IEEE Trans. Ultrason. Ferroelectr. Freq. Control **54**, 836–846 (2007)
25. J.M. Bravo, J.V. Sánchez-Pérez, M. Ferri, J. Redondo, R. Picó1, Application of ultrasound phase-shift analysis to authenticate wooden panel paintings. Sensors **14**, 7992–8002 (2014)

Chapter 6
Case Studies of THz Pulsed TDI

THz pulsed TDI has been used to study various museum objects around the world. This chapter introduces its practical examples, using a commercial THz pulsed TDI system (T-Ray 4000, API Ltd., Ann Arbor, MI, USA). The imaging results presented here should help potential users to understand the benefits and the limitations of THz pulsed TDI for such applications. Additionally, the examples given here may also prompt THz experts to think of new applications or new functions for their THz products and systems.

6.1 Fabric and Mummies

In Sect 4.2, applications to multilayered paper sheets were introduced. THz pulsed TDI can be useful in investigation of multilayered fabric sheets, such as wrapped objects such as mummies, and in the inspection of damage to displayed textile objects. The condition of an original canvas lined on another new canvas for conservation purposes can be examined before additional treatments are performed. Figure 6.1a shows a set of fabric sheets that are laid one on top of another. The piece with the darker colour is oil- and resin-permeated, and the brighter piece below the new linen is oil-permeated. Both fragments are taken from canvas paintings dated c. 1850 CE, and a piece of new linen cloth was placed on top of the two fragments. The observation area is indicated by the white dotted square in the photograph on the left-hand side. The THz reflection image on the right-hand side clearly shows that the new linen cloth is almost transparent to THz waves, and that the darker piece is on top of the brighter piece. Double-layered sheets of the same fabric with and without paint are shown in Fig. 6.1b. The observation area is indicated by the black dotted square in the photograph of the sheet without paint that is placed on the painted sheet. The THz reflection image on the right-hand side shows that the paint itself can clearly be detected through the fabric. This result proves that the canvas

K. Fukunaga, *THz Technology Applied to Cultural Heritage in Practice*, Cultural Heritage Science, DOI 10.1007/978-4-431-55885-9_6

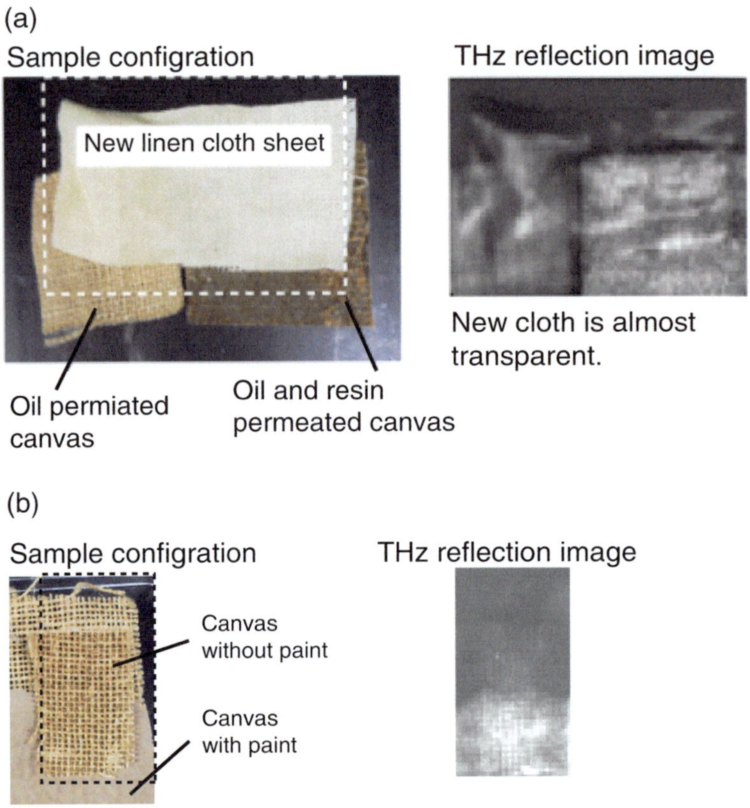

Fig. 6.1 THz pulsed TDI of multilayered fabric sheets from a canvas painting from c. 1850 CE. (**a**) Multi-layered fabric sheets with oil and/or resin content under a new linen sheet, (**b**) double-layered fabric sheets with and without paint

condition of the painting can be examined using THz waves, and Sect. 6.6 introduces suitable examples, including that of a modern painting.

Figure 6.2 shows THz reflection images of multilayered fabric sheets that were originally used to wrap mummies. Conservators in the Metropolitan Museum of Art in New York are interested in using THz pulsed TDI techniques to study a mummy, and the embroidered fabric sheet shown in Fig. 6.2a was provided for trial measurements. After observation of the original sheet, another two fabric sheets (denoted by A and B) were placed on top and the same area was studied again. The embroidered pattern appeared in the THz image, as shown in Fig. 6.2b and the cross-section image along the white line a-a′ indicates that the embroidered part was made from a different sheet and the two were sewn together at the edge with a margin of c. 5 mm. The THz reflection image is formed by the pattern of the threads. Because most dyes are transparent in the THz frequency range (Fig. 3.8), the pattern should be formed by the directions of the stitches in the threads. Additionally, there may also be surface scattering and multiple reflections at the

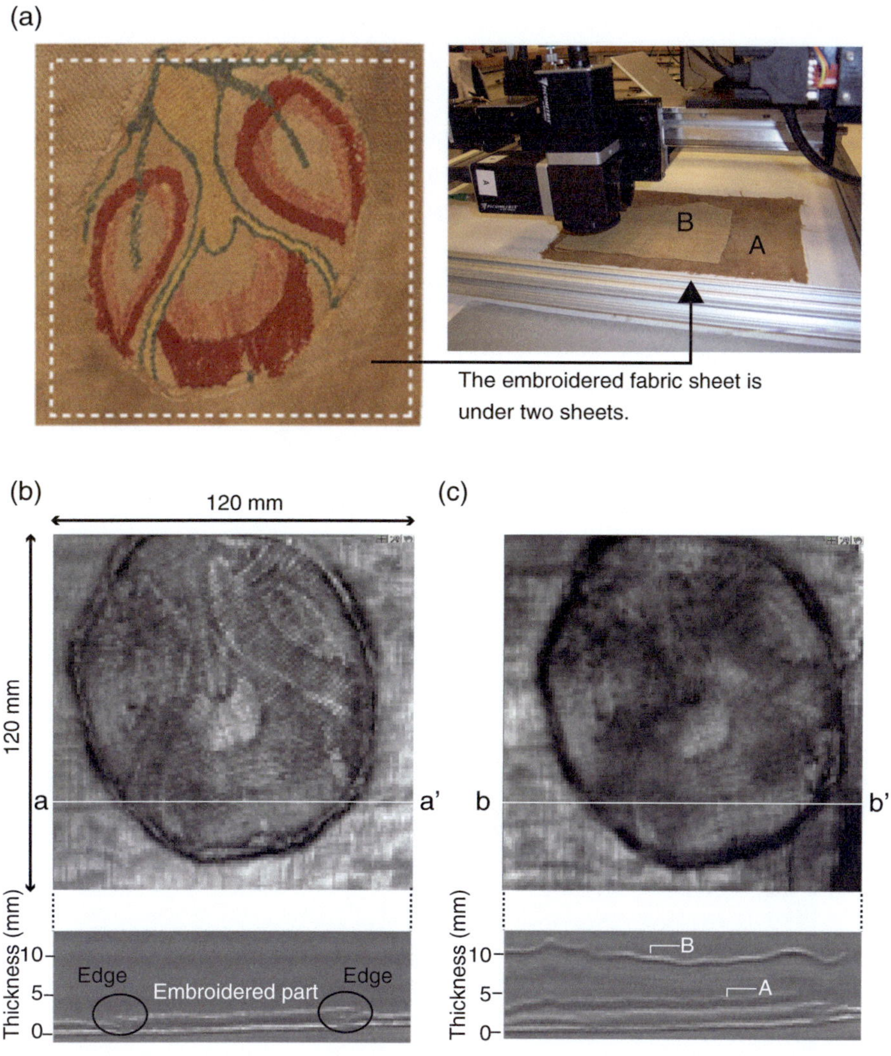

Fig. 6.2 THz pulsed TDI of multilayered fabric sheets from mummies. (**a**) Specimen and measurement setup, (**b**) THz reflection image and cross-section of the embroidered fabric sheet, (**c**) THz reflection image and cross-section of the embroidered fabric covered by two more fabric sheets

surface of the embroidered part. The reflections at each surface of the sheets are strong enough to enable a cross-section image to be acquired, and yet the pattern remains visible through another two sheets, as shown in Fig. 6.2c. These results confirmed that THz pulsed TDI can see through multiple fabric layers.

A mummy in a display was then observed as shown in Fig. 6.3. The mummy of Kharushere (Metropolitan Museum of Art, New York, NY, USA; item id: MMA

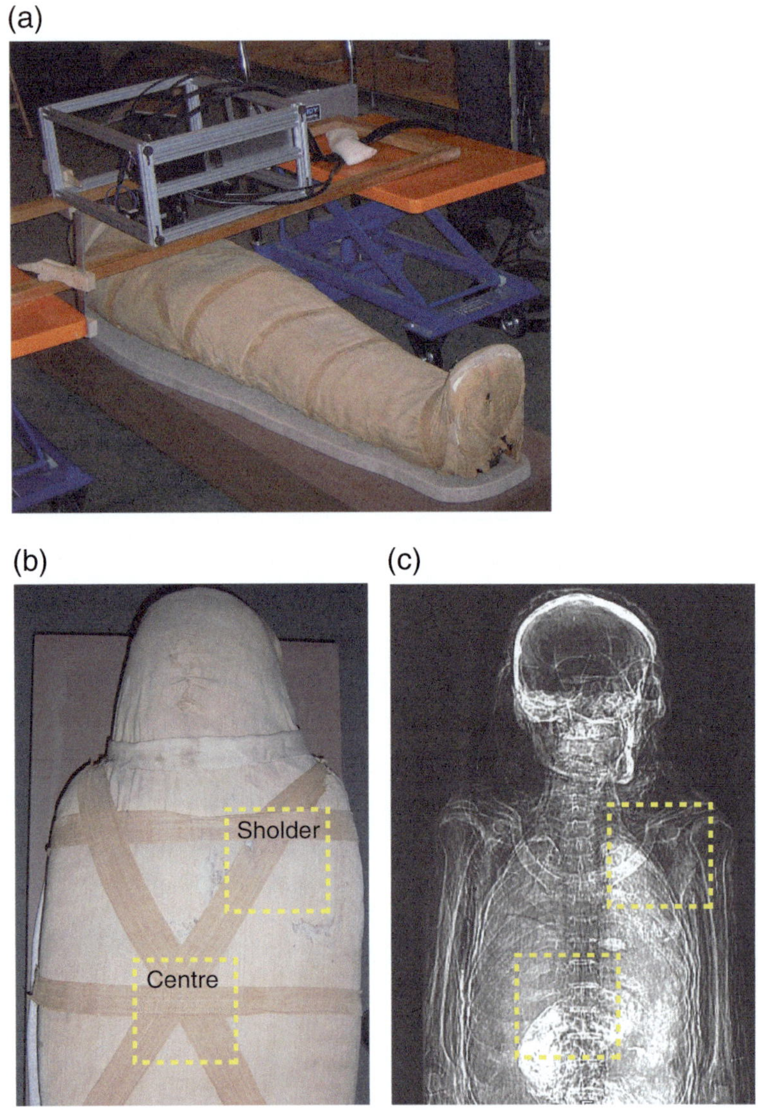

Fig. 6.3 THz pulsed TDI of a wrapped mummy in the Metropolitan Museum of Art, New York. (**a**) Measurement setup, (**b**) observation area, (**c**) X-ray radiography of the observation area from the image obtained by the X-ray CT scanning system at the New York Hospital Department of Radiography

86.1.35; funds from various donors, 1986), who lived during Egypt's 22nd dynasty (c. 825–712 BCE), had previously been investigated by X-ray CT by the museum in collaboration with the Department of Radiography of the New York Hospital [1–3]. This was the first observation of a fully wrapped human mummy by THz pulsed

TDI in the world, and for this study, the T-Ray 4000 system was placed above the mummy, as shown in Fig. 6.3a.

As described earlier in Chap. 4 (Fig. 4.15), the curvature made the measurement more challenging than simple multilayered fabric specimens. Two relatively flat areas (the shoulder and the centre) were chosen for observation, as indicated by the dotted yellow squares in the photograph of Fig. 6.3b, and X-ray radiography (Fig. 6.3c) clearly showed the mummy's body structure.

The THz reflection image of the shoulder area is shown in Fig. 6.4a, and the cross-section along the white broken line a-a′ is shown in Fig. 6.4b [4]. Approximately a quarter of the area on the right-hand side was unclear because of the inclination of the mummy. In addition to the wide rectangular patterns that correspond to the outer bands, there are also lines in the image that are considered to be the bandages below the shroud. The THz image shows the invisible wrapping pattern. The cross-section image clearly shows two very thin layers on the surface, and other sets of layers underneath. The top two layers are obviously the layer of bands, and the internal layer structure indicated the process of mummification. To clarify the number of layers that was present, the output signal waveform at position A in Fig. 6.4a was extracted. The waveform is shown in Fig. 6.4c. The first two large signals that appeared as the top two layers in the cross-section are from the outer bands, and were followed by a small signal from the shroud. The internal bandages were wrapped according to a specific rule to make a set of four layers. The layer structure of the internal bandages of a mummy was observed nondestructively for the first time. However, the mummification process has already been studied thoroughly in a number of previous destructive tests. The interest at the museum is in the mummified human body for investigation of life in the Egyptian era. It is obvious that the THz waves cannot reach the body from the surface, and thus THz pulsed TDI did not provide any new information about the body for historians.

The THz reflection image of the centre part is shown in Fig. 6.4d. The left half was unclear because of the inclination of the mummy. Some irregular objects are detected by comparison with the shape of the shoulder, although these objects are unclear in the X-ray image. For example, a butterfly-like shape appears in the middle of the image. However, it was not a new discovery because it is well known that various objects were inserted between the bandages during the mummification process.

While THz pulsed TDI is unsuitable for investigation of the bodies of wrapped mummies, the results demonstrate that THz pulsed TDI provides rich information about the structures of the multilayered textile works, and about the objects that are covered by the multilayered fabric sheets.

Additionally, there are a number of other excellent works on unwrapped mummies, and a study of a wrapped bird mummy also showed impressive results, including an image of the bird's feet that was invisible to X-ray radiography [5–10].

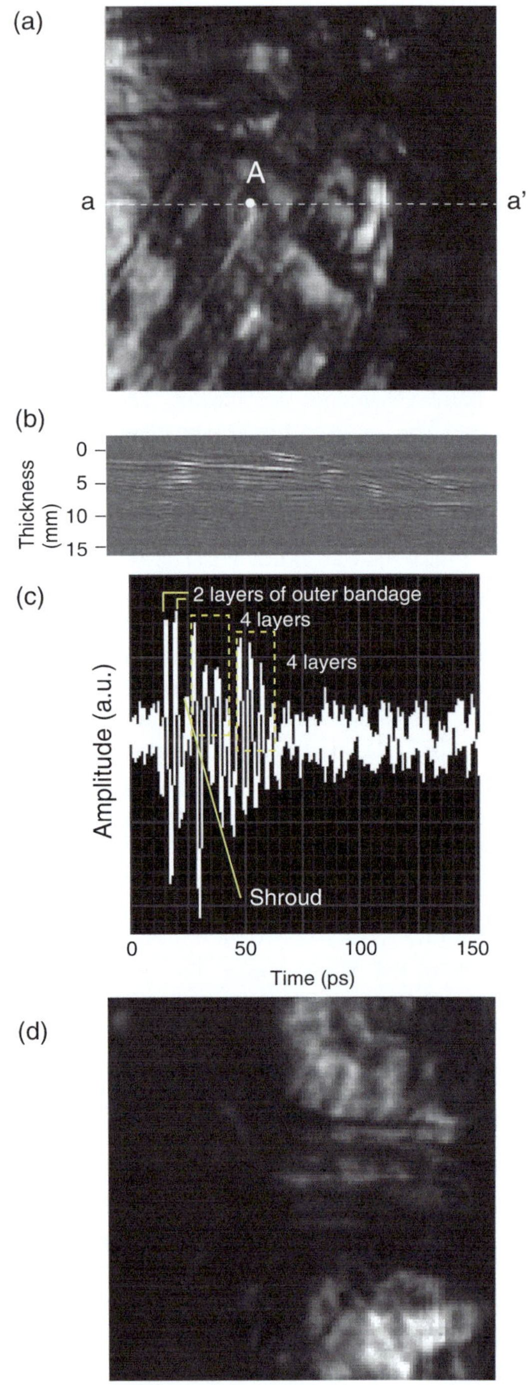

Fig. 6.4 THz pulsed TDI of shoulder and centre parts of mummy. (**a**) Reflection image of shoulder part, (**b**) cross-section image along the *white broken line* a-a′ in part (**a**), (**c**) The output signal at point A indicated in part (**a**), (**d**) THz area image of the centre area

6.2 Wall Paintings

Wall paintings span one of the longest periods of time in cultural heritage, from the days of the Palaeolithic era up to the present day, and occur globally. Exploring wall paintings from all time periods is extremely important for an understanding the history of the painting activities of humanity. There have been various attempts to apply THz pulsed TDI to wall painting analysis, where the main objective of these previous works was to discover under-drawing or under-painting [11, 12]. In this chapter, in contrast, THz pulsed TDI is applied to investigation of the structural information of wall paintings, which is directly related to the painting technique, and to examination of the preparation layer, which is important for practical conservation planning.

Among wall painting methods, "buon fresco" painting is currently regarded as a major technique. It is a wet painting process developed or refined by Giotto di Bondone around 1300 CE [13]. In this technique, the pigments are painted on a wet mortar preparation layer with water only. The pigments become fixed in the wall as the surface is covered by a crystalline calcite layer that is generated by the chemical reactions among the mortar, the water, and carbon dioxide in the air. Other European wall paintings, e.g., Pompeian [14], Etruscan, and Byzantine paintings, used a similar wet painting process. The differences among these techniques are in the layer structure used for preparation and the aggregates used in the mortar. It is normal that the wall painting material varies, because natural resources for paintings, including minerals and the environmental conditions, depend on the location.

Another painting method based on mortar preparation is "fresco a secco", a dry processing technique that uses a binder to fix the pigments to the dried wall. This technique is common in Asia, and is also used for conservation applications in Europe. The aggregate in the mortar layer in Asia is usually composed of clay and fibres made from straw, linen, or bast, while animal glues are used as the binders.

In contrast, for the paintings in Egypt, the wall was not made from mortar. The walls were made from gesso and clay, then gum arabic is used as the binder required to paint. Egyptian painting can thus be categorised as watercolour painting.

Understanding of the layer structure of a wall painting can help us understand the history of its manufacture and the possible cultural exchanges involved in the process. To investigate the possible use of THz pulsed TDI for wall paintings, models prepared using eight traditional techniques were studied. The models, which are shown in Fig. 6.5, were made by Misao Ohno, and were displayed in an exhibition at Tama Art University in 2009.

The THz pulsed TDI results for the Pompeian model shown in Fig. 6.5a were introduced in Chap. 5 as part of a comparison between NMR and THz pulsed TDI techniques. Another seven techniques, which are shown in Fig. 6.5b–h, are discussed in this chapter. The THz pulsed TDI experiments were performed using preparation layer models of the seven techniques, and these models were also observed by NMR, as described in Sect. 5.1 [15].

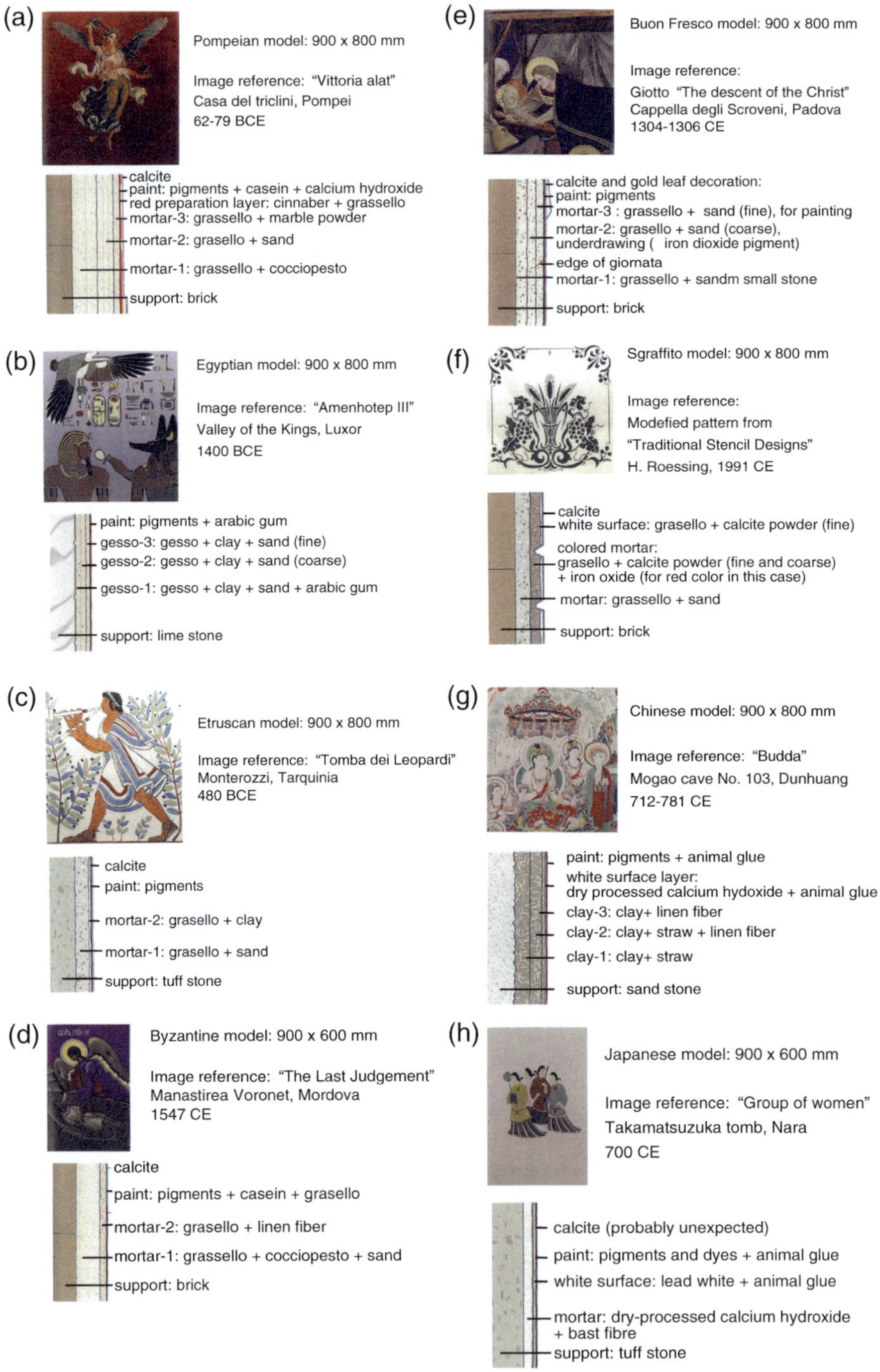

Fig. 6.5 Eight examples of classical wall painting techniques from around the world. (**a**) Pompeian, (**b**) Egyptian, (**c**) Etruscan, (**d**) Byzantine, (**e**) buon (Florentine), (**f**) sgraffito, (**g**) Chinese (Dunhuang), and (**h**) Japanese

Figure 6.6a–d shows photographs of the models in top and tilt views, and the THz cross-section images. A typical output signal waveform is superimposed as a yellow line on each of the cross-section images of four wet-process models of the Etruscan, Byzantine, buon, and sgraffito techniques. The layer structure of the Etruscan model shown in Fig. 6.6a is unclear. This is because the resolution in the thickness direction was not fine enough to distinguish between the two extremely thin mortar layers. Additionally, this model has multiple surface spots, which cause scattering, and multiple reflections, which result in reduced THz wave propagation in the layer. In the Byzantine model shown in Fig. 6.6b, the interface between the paint layer and the top mortar layer can be recognised. This model also has a few surface spots, and a small separation was found. The layer structure of the buon fresco model can be clearly seen, from the paint to the surface of the second mortar layer, as shown in Fig. 6.6c. It is also confirmed that the top mortar layer was made in two-ply form by using the same material twice. A fresco wall made by the sgraffito technique is shown in Fig. 6.6d. In this technique, the designed coloured mortar pattern is actually created by scratching it into the white surface layer. The layer structure from the surface to the preparation materials can be seen. It is also clear that each mortar layer was made in two-ply form by using the same material twice.

Figure 6.7a–c show the photographs in the top and tilt views, along with THz cross-section images with a typical output signal superimposed on the cross-section image, of the dry process wall painting models. In Egyptian wall painting, shown in Fig. 6.7a, the layer is only visible at the first interface, and no signal follows. This is because the relatively coarse sand aggregates generate internal scattering, and prevent THz wave propagation. The preparation layer for the Chinese (Dunhuang) technique shown in Fig. 6.7b is made using clay with straw aggregates. The interface between the clay layers can be recognised, although the layer thickness varies. When compared with the other wall paintings, the preparation layer of the Japanese wall painting is extremely thin, as shown in Fig. 6.7c. Thus the THz wave can propagate through the mortar layer and can thus reach the stone support. The internal interface of the mortar layer appeared clearly in the image and anchor scratches on the support stone were recognised.

Also, in terms of depth profiling, no interfaces were found in the Egyptian and Chinese models by the NMR depth profile method [15]. This is because the water disperses homogeneously during the measurements.

Because the Japanese wall painting has a thin preparation layer with a thickness of approximately 5 mm or less, and because no coarse aggregates were used, THz pulsed TDI is suitable for observation of the internal condition of the painting, from the surface to the supporting stone. It should be noted here that only three wall paintings with mortar have been found in Japan to date, in contrast to those in continental countries [16]. Other more common wall paintings were drawn directly on the wall, and the figures are more like geometrical patterns. The mortar preparation layers of each of these three paintings are extremely thin when compared with continental wall paintings. The reason is yet unknown.

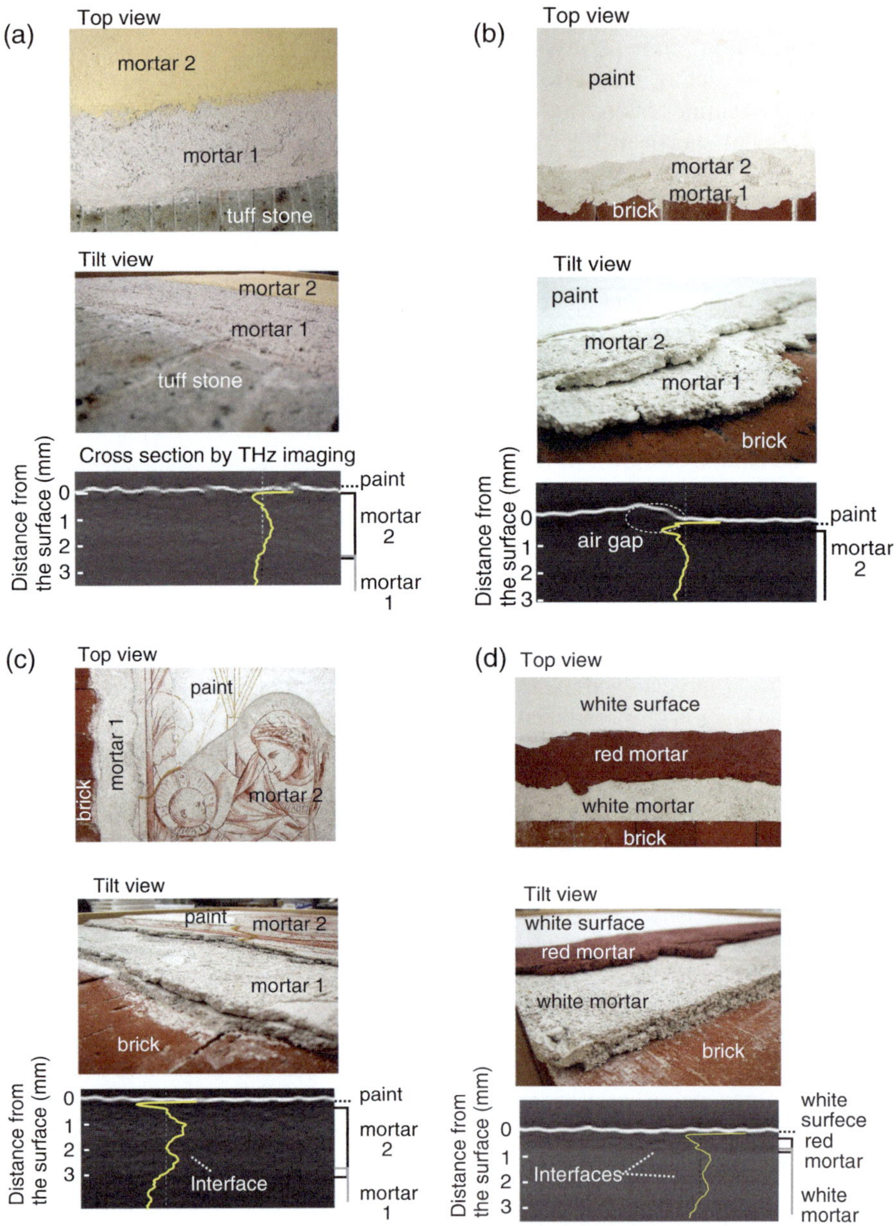

Fig. 6.6 Preparation layer models prepared by wet processes and their THz cross-section images. (**a**) Etruscan, (**b**) Byzantine, (**c**) buon/Fiorentino, and (**d**) sgraffito

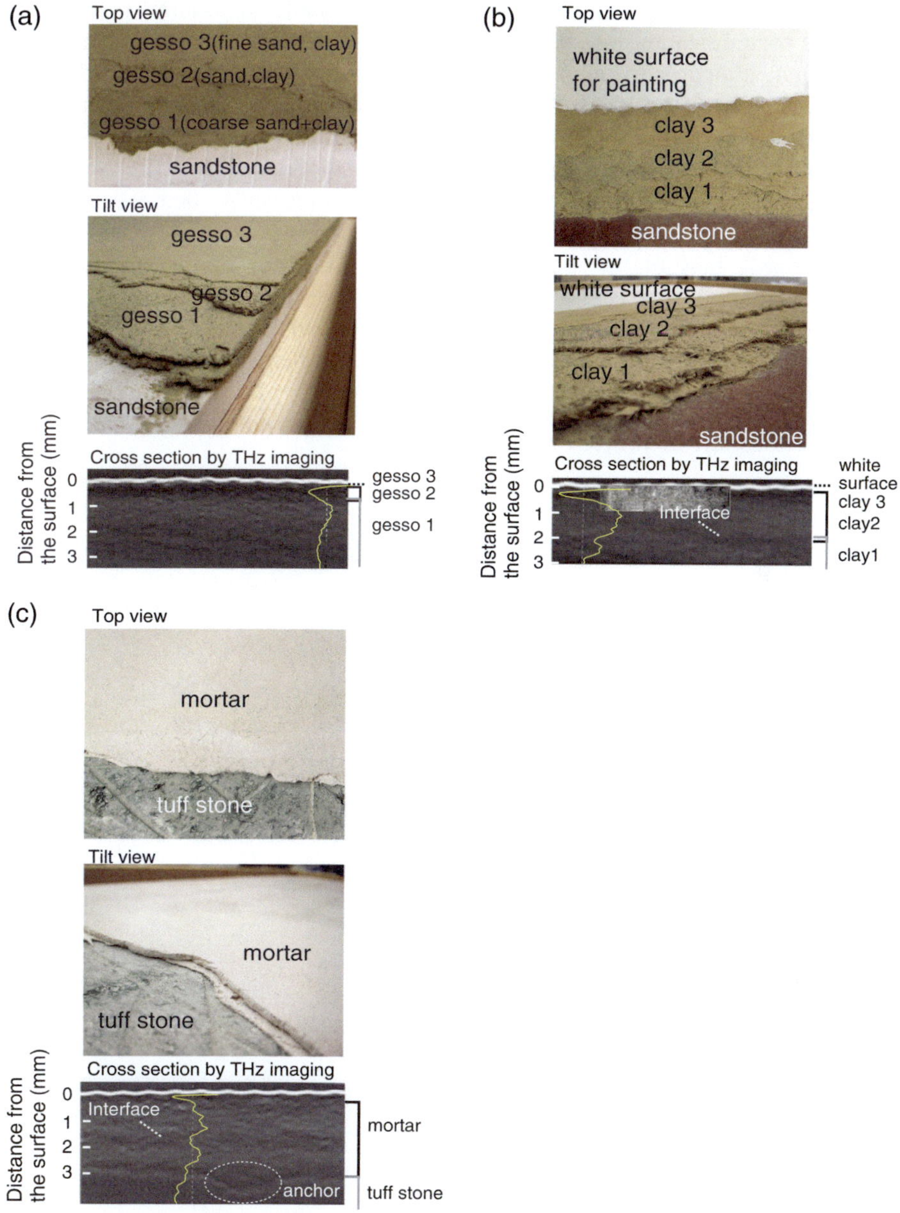

Fig. 6.7 Preparation layer models prepared by dry processes, and their THz cross section images. (**a**) Egyptian, (**b**) Chinese (Dunfuang), (**c**) Japanese

All three wall paintings, i.e., Takamatsuzuka, Kitora, and Marukoyama, were found in Nara Prefecture, where the first governmental capital was established in 710 CE. The city was designed in the Chinese style, and the advanced continental culture then rushed into the city. It is thus reasonable to assume that a new technique for fabrication of a wall painting with mortar was introduced to Japan from China, although Chinese wall paintings tend to have thick multilayered preparation layers. The objects that were painted in the Takamatsuzuka and Kitora tumuli are also based on a philosophical idea that was popular in China at that time [17], while the Marukoyama has only one white mortar layer. The tumuli of Takamatsuzuka and Kitora and their current states after severe deterioration [18] are introduced in Fig. 6.8.

The overall view of the tombs before the disassembly of the stones in the Takamatsuzuka tumulus is shown in Fig. 6.8a. This photograph was taken from the south, where grave robbers in the twelfth century had made a hole in the wall to gain entry. While the wall paintings had vivid colours when they were discovered in 1972, they had suffered from heavy deterioration because of a number of factors. In 2007, the stones were disassembled, and they are now stored in a conservation

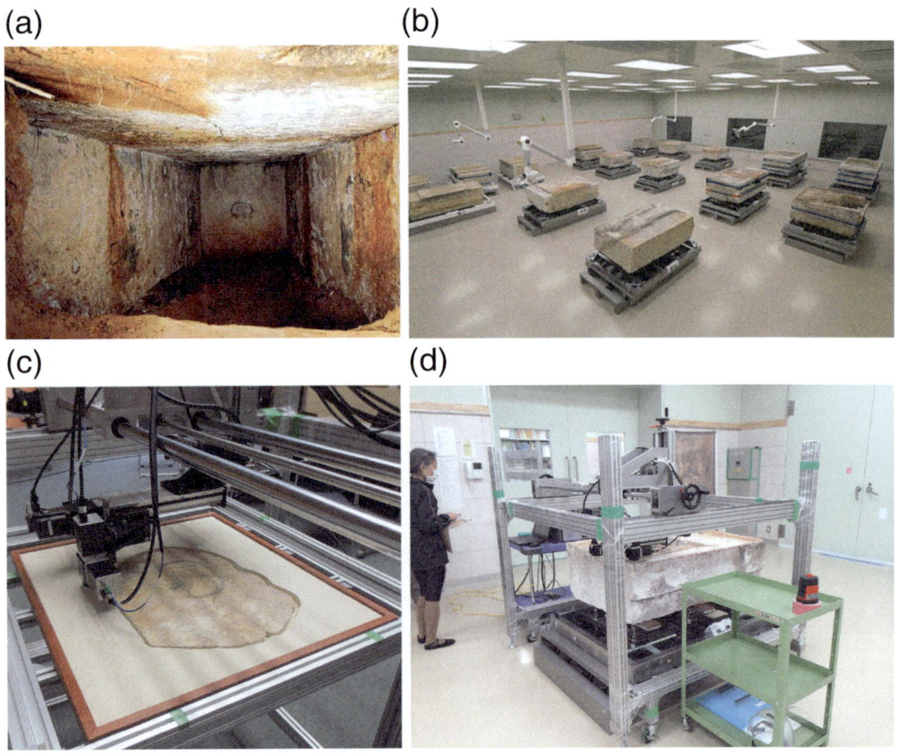

Fig. 6.8 THz pulsed TDI measurement setup for the Takamatsuzuka and Kitora tumuli. (**a**) Photograph of Takamatsuzuka tomb before disassembly in 2007, (**b**) disassembled stones with wall painting from Takamatsuzuka tomb, (**c**) measurement setup used for wall paintings that were removed from the wall in the Kitora tomb, (**d**) measurement setup used for the wall paintings on the disassembled stones from the Takamatsuzuka tomb. All visible images in this section were taken by Nara National Research Institute for Cultural Properties

facility where the environmental conditions are strictly controlled, as shown in Fig. 6.8b. In contrast, the Kitora tumulus was discovered in 1983, and all painted areas were removed from the preparation layer and were examined by various methods, including infrared imaging. Figure 6.8c, d show the THz imaging setup used for the wall paintings from the Kitora and Takamatsuzuka tombs. The position of the sensor unit, which is equipped with an automatic X-Y position controller for an area of $180 \times 180\ mm^2$, can be manually moved above the painting within the custom-built frame.

Figures 6.9, 6.10, 6.11, and 6.12 show the THz reflection observations of four gods that were painted in the four directions of the Kitora tomb, based on Chinese mythology. These paintings were removed from the wall from the preparation layer and were placed on honeycomb structural plastic foam boards (Rohacell®). .

Figure 6.9 shows the THz reflection observation of the "Genbu (Black Tortoise)" painting on the north wall of the Kitora tomb. The snake and the turtle both remain in good condition, although their colours are not bright, as shown in Fig. 6.9a. The snake shape is recognised as an oval dark area in the middle of the THz area image shown in Fig. 6.9b. The most striking feature of the image is generated by the condition of the preparation layer. Surface cracks and material losses are clearly observed. While the snake and the turtle on the surface remain presentable, the cross-section shown in Fig. 6.9c revealed that the preparation layer of Genbu is entirely porous, i.e., it is heavily deteriorated.

The God of the East, Seiryu, is barely visible as the Blue Dragon in Fig. 6.10a. Clay, mud and water from the top of the wall had covered the figure. While the THz area image shows some large cracks, the reduced number of internal reflections proved that most of the preparation layer is relatively in good condition, i.e., homogeneous and solid, as shown in Fig. 6.10b. Cross-section images shown in Fig. 6.10c along with the red dotted line and 6.10d along with the yellow dotted line indicated that some parts of the surface that were coated with clay seemed to be lifted, while at the actual surface, some parts had a new layer that can be generated via treatment with hydroxyl propyl cellulose (HPC), which also appeared on the south wall.

The shape of the Red Phoenix "Suzaku", of which a carbon black drawing is shown in Fig. 6.11a, is clear and the red colour with gradation appears extremely vivid. The THz reflection image shows some evidence of an engraving underdrawing around parts of the tail, which are not exactly traced by the drawing, on the left-hand side of Fig. 6.11b. This indicates that the underdrawing is like a rough sketch that is used to decide the size and position of the phoenix. The engraving underdrawing generally appears as dark lines in the THz reflection image due to scattering at its edges, resulting in low reflection levels.

In the middle and on the right-hand side of Fig. 6.11b, relatively wide white perpendicular lines are observed. These lines were deduced to have been generated by the HPC surface treatment. The cross-section images along the magenta line are shown in Fig. 6.11b and those along the perpendicular line and along the middle white line are shown in Fig. 6.11c, d, respectively. The greatest advantage of THz TDI is that the area image of the layer of interest can be extracted from the data, as

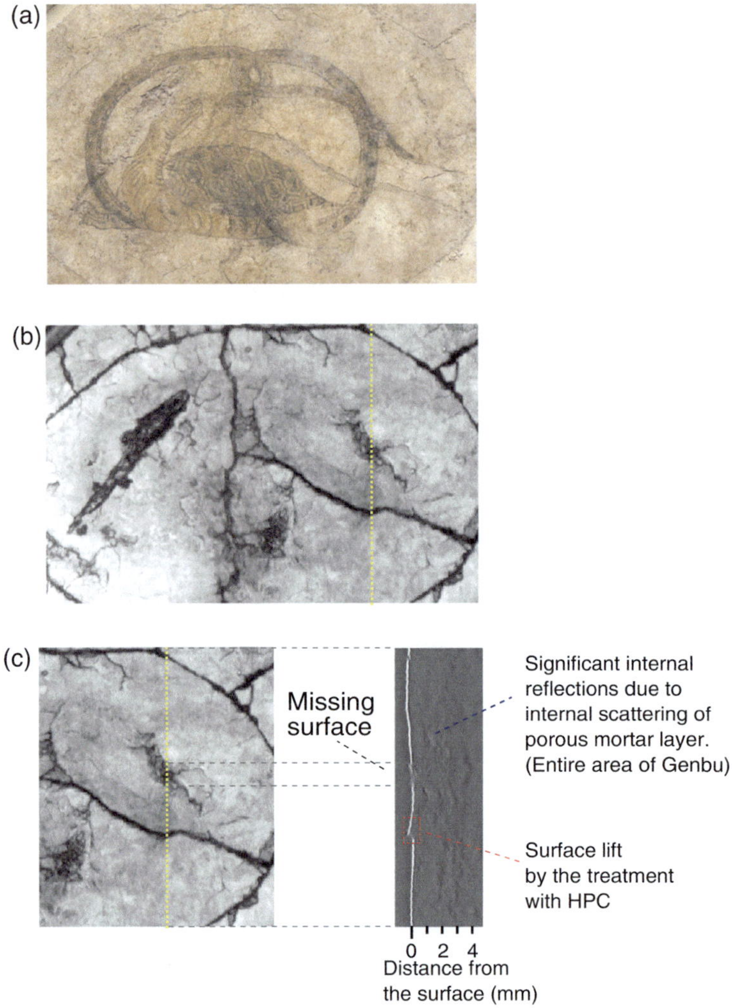

Fig. 6.9 THz reflection observation of "Genbu" on the north wall of the Kitora tomb. (**a**) Visible image, (**b**) THz area image, (**c**) cross-section image along the *yellow dotted line* shown in part (**b**)

described in the previous chapters. This means that it is possible to observe slices of the painting nondestructively by peeling from the surface. Figure 6.11d is the area image at approximately 1 mm below the surface. A peculiarly strong reflection obtained in this area indicated the presence of an air void. According to the conservation scientists, ethanol was used for cleaning, then acetic acid bacteria consumed the ethanol and generated the acid. The acid causes the mortar layer to deteriorate, resulting in air gaps [19]. These results proved that THz TDI is useful for observation of original artworks and for conservation processes.

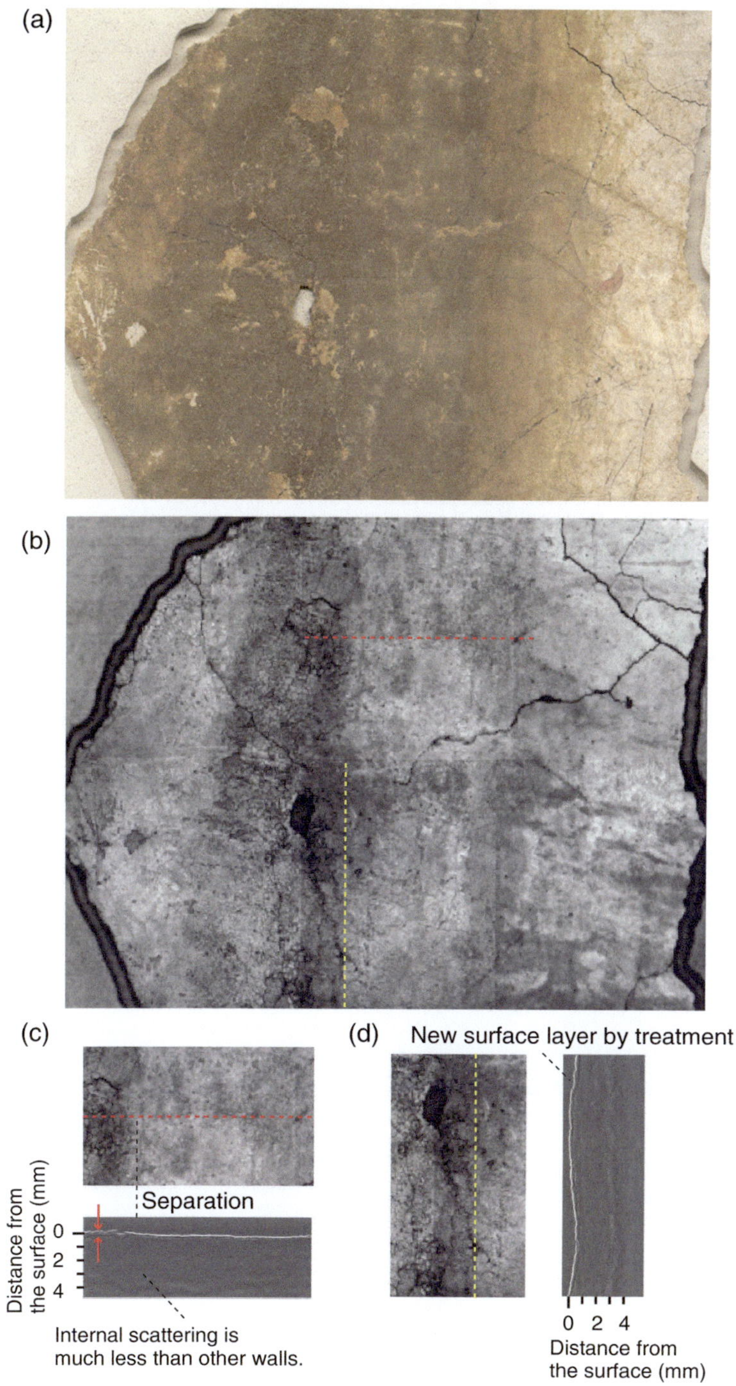

Fig. 6.10 THz reflection observation of "Seiryu" on the east wall of the Kitora tomb. (**a**) Visible image, (**b**) THz area image, (**c**) cross-section image along the *red dotted line* shown in part (**b**), (**d**) cross-section image along the *yellow dotted line* shown in part (**b**)

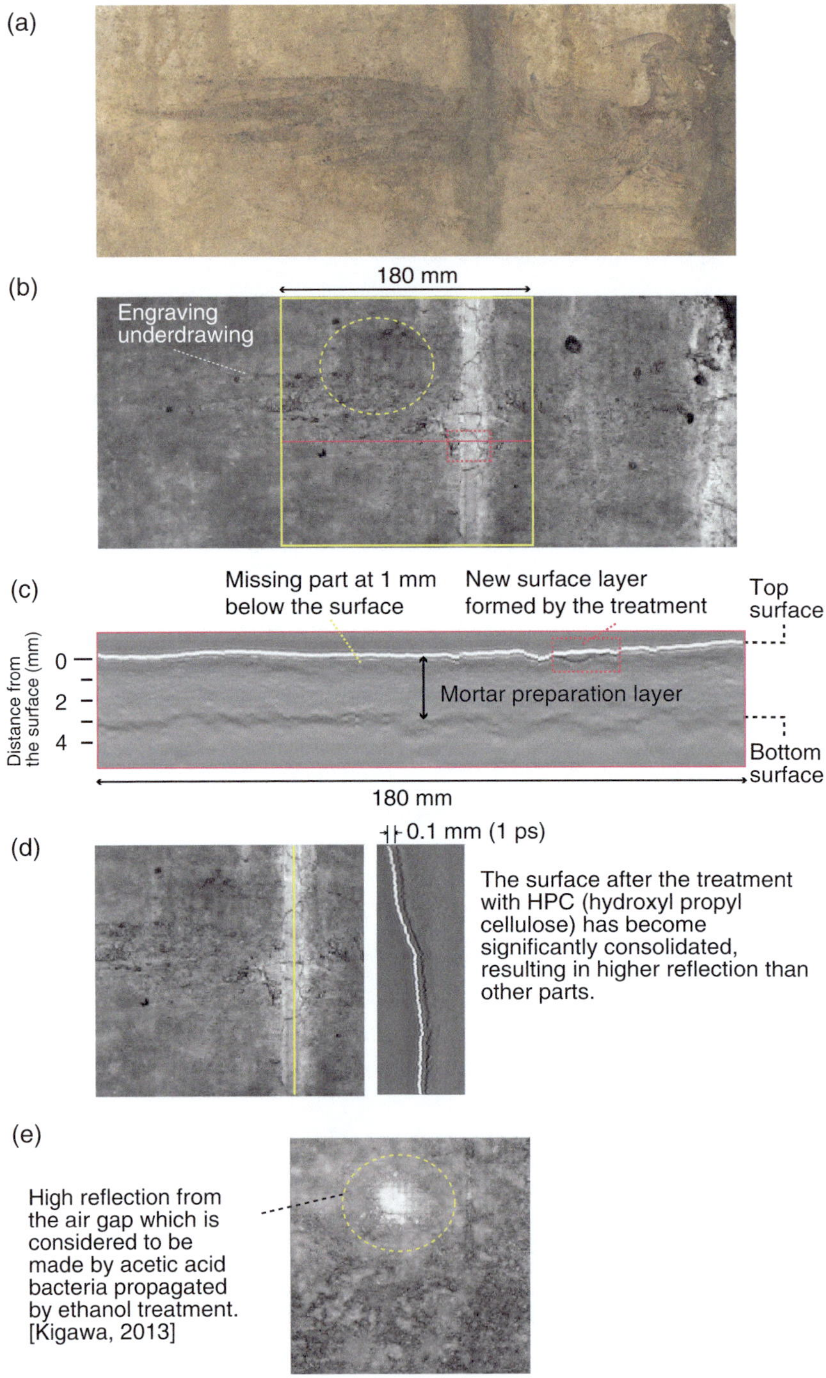

Fig. 6.11 THz reflection observation of "Suzaku" on the south wall of the Kitora tomb. (**a**) Visible image, (**b**) THz area image, (**c**) cross-section image along the magenta line shown in part (**b**).

The God of the West "Byakko" is as shown in Fig. 6.12. Because no wild tigers existed in Japan, the figure was designed based on the artist's imagination, combined with some information that came from the continent. Figure 6.12b shows the THz area image of the entire fragment. Around the front legs, which are indicated by the yellow circle, the engraving underdrawing is clearly visible. Cross-section images along the yellow and red dotted lines are shown in Fig. 6.12c, along with the area image that was extracted from Fig. 6.12b. The preparation layer of the area, which has the engraving underdrawing, is in good condition, based on the fact that almost no internal reflection signals were observed. However, the preparation layer of the fragment located next to this area was heavily deteriorated. The preparation layer along the yellow dotted line shows a double layer structure, which indicates that the clay on the surface formed a new surface layer through the treatment with HPC, as described previously in other walls [16].

The Takamatsuzuka wall paintings were found in 1972 by chance, and their vivid colours and precise drawings attracted considerable public attention. Unfortunately, over time, these wall paintings have deteriorated heavily. Some parts of the paintings are covered with fungi, and some of the colours have diminished.

While the conservation techniques used for common Japanese paintings are well established, based on a rich history and knowledge of the subject, the techniques required for wall paintings are still under development. This is because most common Japanese heritage objects are made from paper and glue, and the mortar-like layers only exist at the very surface to form relief-like features, as will be described later in Sect. 6.5. Thus, the conservation of the wall itself was not performed properly within the tombs, and the government finally decided to disassemble the stones. The Nara National Research Institute for Cultural Properties conducted the project, and in 2007, all stones were finally placed in a special room where the environment is strictly controlled. The current status of the wall is that the surface has dried out and the black spots that were generated by fungi have been partially removed. Scientific examinations have continued, including THz pulsed TDI studies for preparation layer observation.

The most famous part of the Takamatsuzuka wall painting is the "Ladies on the West Wall". Fig. 6.13a shows a photograph of the observed stone surface, before the disassembly of the tomb. Because the THz pulsed TDI system used in this work can observe an area of 180 × 180 mm in a single scan, the THz reflection image shown in Fig. 6.13b is composed of 35 images. THz pulsed TDI observation is usually performed at a resolution of 1 mm or less because the resolution can theoretically reach approximately 0.2 mm. However, this entire stone was observed at a resolution of 2 mm, except for two areas, due to the limitations of the observation time. The grey scale is automatically set in each square image, and

Fig. 6.11 (continued) (**d**) HPC-treated area and enlarged cross-section image along the *yellow solid line*, (**e**) Area image of the *yellow square* shown in part (**b**) at a depth of 1 mm from the surface

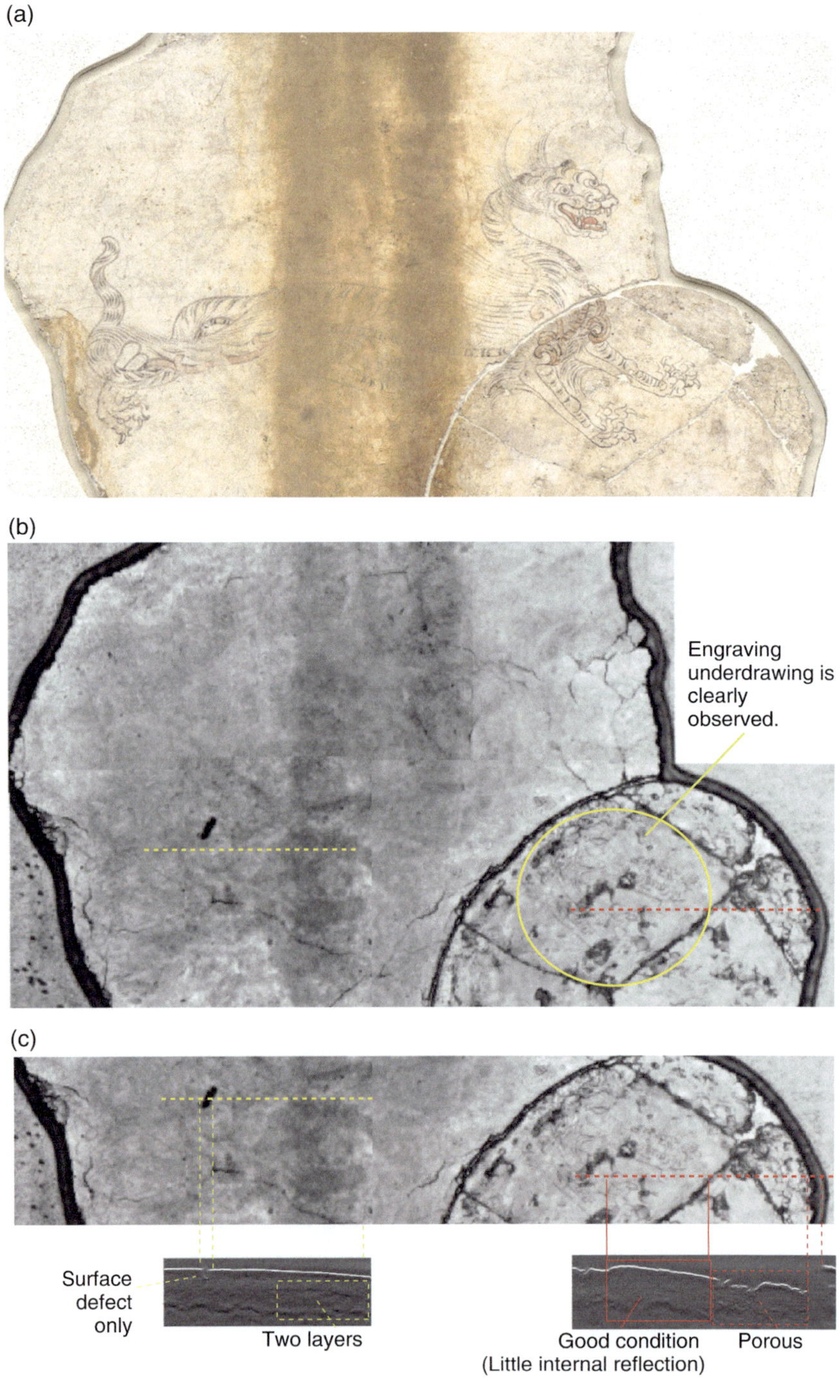

Fig. 6.12 THz reflection observation of "Byakko" on the west wall of the Kitora tomb. (**a**) Visible image, (**b**) THz area image, (**c**) cross-section images along the *yellow* and *red dotted lines* in (**b**)

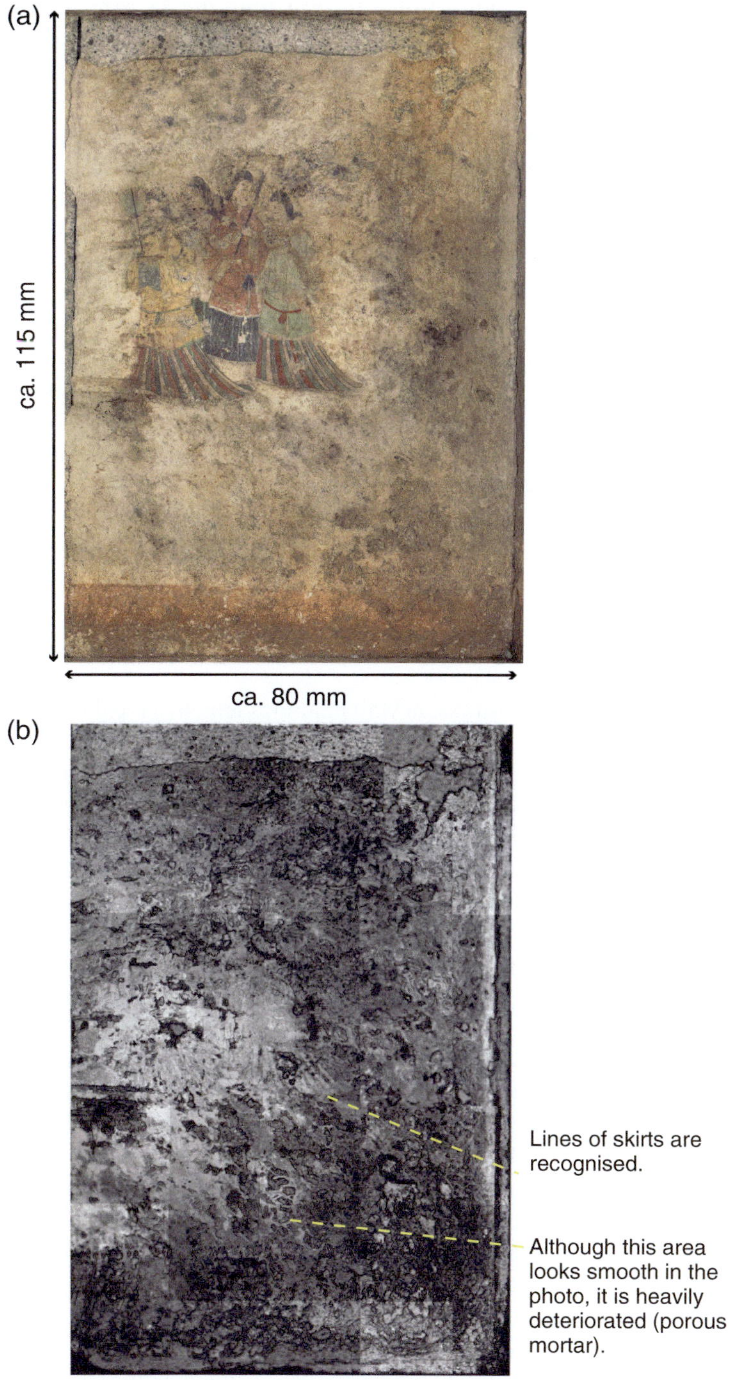

Fig. 6.13 THz reflection imaging of the north stone of the west wall in the Takamatsuzuka tumulus. (**a**) Photograph taken before the disassembly of the stones, (**b**) THz reflection image of the entire area

thus the grey scale over the entire stone, formed from the 35 tiled images, is not quantitatively corrected.

As shown in this figure, the women that were painted on the surface were hardly recognised by THz pulsed TDI, except for some of the lines of their skirts, because the painting layer itself is thin and contains only small amounts of pigments, and the lines have been painted with carbon black, which is transparent in the THz frequency region. The THz image in Fig. 6.13b showed the inhomogeneous features of the preparation layer. The entire preparation layer is porous, and the degree of deterioration does not appear directly on the surface.

During the measurements, the conservators realised that some of the THz reflections corresponded to the areas where previous conservation treatments were conducted. Some parts of the preparation layer were previously consolidated with acrylic resin. Although these areas are invisible, they appeared as dark spots in the THz reflection images, as shown in Fig. 6.14a, b. The distribution of dark spots that appeared in the smooth area is in good agreement with the consolidated parts that are indicated by the shaded portions shown in Fig. 6.14c. The cross-section image along the dotted line in Fig. 6.14b, as shown in Fig. 6.14d, revealed that the preparation layer that had been treated with acrylic resin is homogeneous, and has adhered well to the stone surface because reflections from the stone surface cannot be detected clearly. In contrast, the area that was not treated shows separation between the preparation layer and the stone because the strong relfection from the stone was ovserved.

Chinese and Continental East Asian wall paintings tend to have thick preparation layers with coarse fibre aggregates, as shown in Fig. 6.7. The Nara National Research Institute for Cultural Properties conducted a collaborative research project with The Inner Mongolia Museum and the China National Institute of Cultural Properties on the wall paintings of the Dazhao Monastery. This Monastery is the oldest Lamaist Buddhist temple in Inner Mongolia; it was built during the Ming Dynasty, c. 1580 CE, and is well preserved. A fragment from the wall painting was once analysed using an X-ray fluorescence spectrometer and a laser Raman analyser at 785 nm, and it was also analysed by THz pulsed TDI in 2010 [20, 21]. Figure 6.15a shows the Raman spectroscopy observation points that are indicated on the photograph as the letters a to e. The orange colour in the white dotted square was not analysed at that time. The pigments that were identified using Raman spectroscopy are shown in Fig. 6.15. The vivid orange pigment of the skirt-like part is minium, and the red pigment is cinnabar. Both pigments have relatively high reflection levels for THz pulsed TDI. The THz reflection image in Fig. 6.15c shows that the orange part that was painted with minium is more reflective than the red part. However, the orange belt in the white dotted square unexpectedly failed to appear in the THz reflection image, while similarly orange-coloured armlets and bracelets were clearly detected. Therefore, XRF element mapping was performed to investigate this aspect of the orange colour. The orange parts that have high reflection characteristics in the THz image were painted with a lead-based pigment, which is most likely minium. The red part was painted with a mercury-based pigment, i.e., cinnabar. In the white dotted square, only arsenic-based pigments,

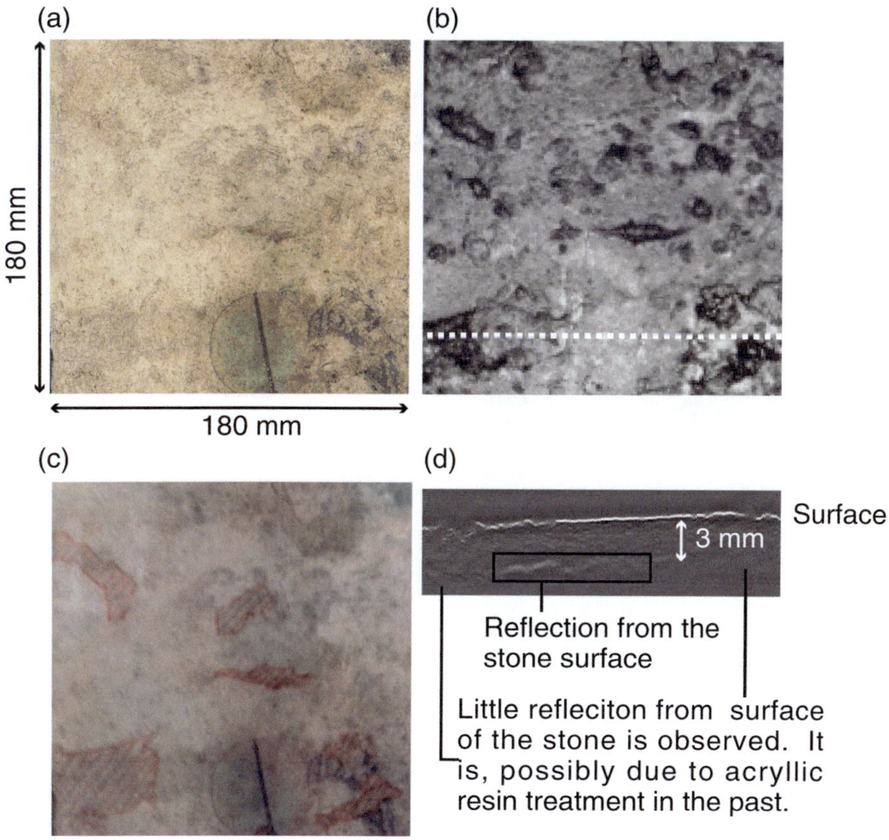

Fig. 6.14 Comparison of THz reflection images with previous treatment map. (**a**) Area under analysis, (**b**) THz reflection image, (**c**) map of previous acrylic treatment recorded by conservators, (**d**) cross-section image along the *dotted line* shown in Fig. 1b

such as orpiment, were found, and no specific mineral was found on the orange belt. Thus, the orange colour could have been painted using an organic dye. The NIR reflection image shown in Fig. 6.15e clearly shows lines drawn using carbon black pigments and some lines on the body that are not visible in the THz reflection image. However, the material of the orange belt has not only been identified by nondestructive methods; the other results suggested the possible use of organic dyes.

The cross-section image produced by THz pulsed TDI was compared with the photograph that was taken at the fringe, which is expected to be similar to the cross-section, and these images are shown as a combined figure in Fig. 6.16. The cross section along the red line in the white dotted square, which is also visible in the THz reflection image on the right-hand side, shows three layers and the effects of some surface cracks. The painting layer and the top preparation layer, which were made with fine pigments, cannot be distinguished. The surface cracks cause scattering at

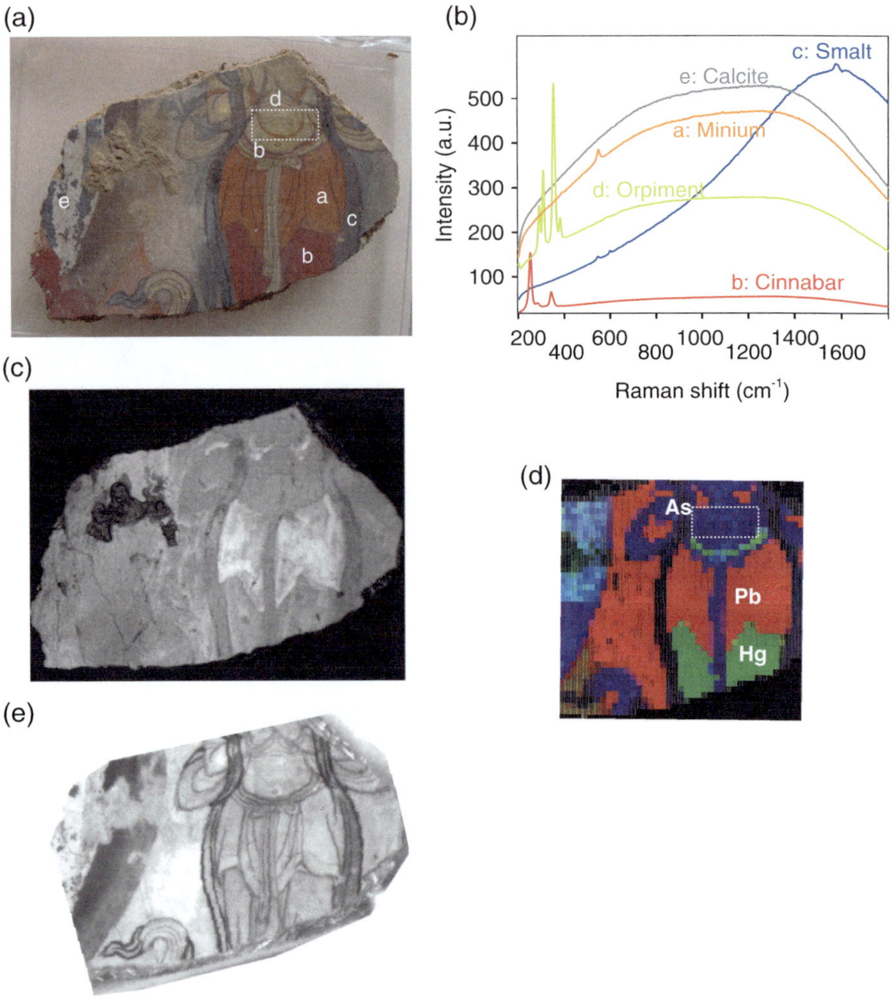

Fig. 6.15 Nondestructive analysis results of a fragment of a wall painting from the Danzhao Monastery in Inner Mongolia, c. 1850. (**a**) Measurement points determined by Raman spectroscopy, as indicated on the photograph, (**b**) pigment identification based on Raman spectroscopy, (**c**) THz reflection image, (**d**) XRF element analysis map performed by Nara National Research Institute for Cultural Properties, (**e**) NIR reflection image

the surface, and this affects the cross-section imaging process. Unlike the case of Japanese wall paintings, THz waves cannot pass through the entire preparation layer, which means that the interface between the preparation layer and the supporting wall can hardly be seen in the THz image.

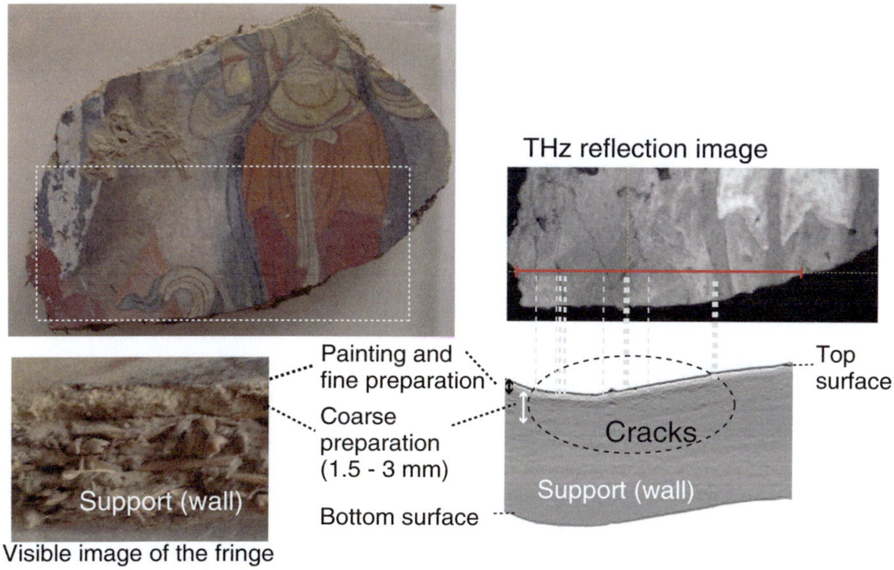

Fig. 6.16 Cross-section observation by THz pulsed TDI compared with the photograph that was taken at the fringe

6.3 Western Panel Paintings

THz waves can reach the wooden supports of panel paintings through the gesso preparation layers; THz pulsed TDI can thus provide nondestructive cross-section images, and if required, area images of the layers of interest, by extracting reflection pulses in the time domain, as illustrated in Fig. 4.3. The cross-section images are useful in practice for conservators for understanding details of the techniques used in the original artwork and in any previous treatments. Additionally, previous works have also proved that THz technology is useful for wood examination [22, 23]. Applications of THz pulsed TDI to the examination of tempera panel paintings on wood have shown successful results [24, 25]. The following examples based on two masterpieces demonstrate how THz pulsed TDI works for panel painting analysis.

The first museum object to be observed by THz pulsed TDI is the "Polittico di Badia" (c. 1300 CE, 334 × 91 cm) by Giotto, from the permanent collection of the Uffizi Gallery in Florence, Italy, as shown in Fig. 6.17 [26]. The masterpiece was in conservation in 2008, and scientific examinations were done at that time.

A comparison of the THz reflection images with the photographs of the saints and an angel is shown in Fig. 6.18. Because metal flat surfaces reflect THz waves almost completely, the reflection levels of the gold leaf areas become the highest in each observation area. In the areas where the gilded parts are not visible, or are even hidden by paint layers, the presence of the gold leaf appears as a high reflection level, as shown in Fig. 6.18a–c. The gold leaf of San Nicola's glove, shown in

Fig. 6.17 THz pulsed TDI of "Polittico di Badia" by Giotto. (**a**) Photograph and observation area by THz TDI, (**b**) measurement setup. Photographs in this section were taken by Antonio Quattrone

Fig. 6.18a, was considered to be used as the base for the velatura with semi-transparent pigments. The gold decoration at the cuff of the sleeve of San Pietro (Fig. 6.18b) is detected, although it appears to be brown in the photograph. This decoration was supposed to be made by the "a missione" technique. The shape of the invisible gold leaf under the paint appears as shown in the angel's face in Fig. 6.18c. Detection of gold leaf or any other type of metal under paint layers is too easy example for THz experts for it to be considered as a serious application of the method. In practice, however, it is helpful for conservators to be aware of fragile gold leaf beneath the paint when they are performing cleaning processes on the surface. In addition to gold, lead white also gives high reflection results in the highlight of the navy mantle and the pen of San Giovanni, as shown in Fig. 6.18d.

Figure 6.19 shows the THz pulsed TDI of the panel of the Madonna with the Child. Similar to the examples shown in Fig. 6.18c, the shapes of the areas of gold leaf under the paint can be clearly observed around the Child's face in the area image near the surface, as shown in Fig. 6.19a. In the same image, part of the hair near the face line of the Madonna was badly deteriorated. This deterioration was visible before the conservation process; however, even if a conservation treatment can cover this part of the panel, THz pulsed TDI can still reveal the condition of the original artwork by detecting the interface between the original and covered layers. Figure 6.19b shows the THz cross-section image along the broken line a-a′. Unlike modern tempera panel paintings, which use canvas or mesh cloths applied directly on the panel, there are two gesso layers in this panel, and the cloth is placed between them. This cross-section image proved that Giotto followed a medieval procedure in making this panel painting. At the time of painting (around 1300 CE), flat wood panels were not commonly available, and a wood panel painting was simply part of the decorative area of a church. According to the conservators, this technique involves a gesso layer being spread directly on the carved wood base to obtain a reasonably flat surface. Then, a cloth was placed on top to ease the tension on the wood. Finally, another gesso layer was formed as a preparation layer for the painting. The area images of each of the layers are shown in Fig. 6.19c. The THz

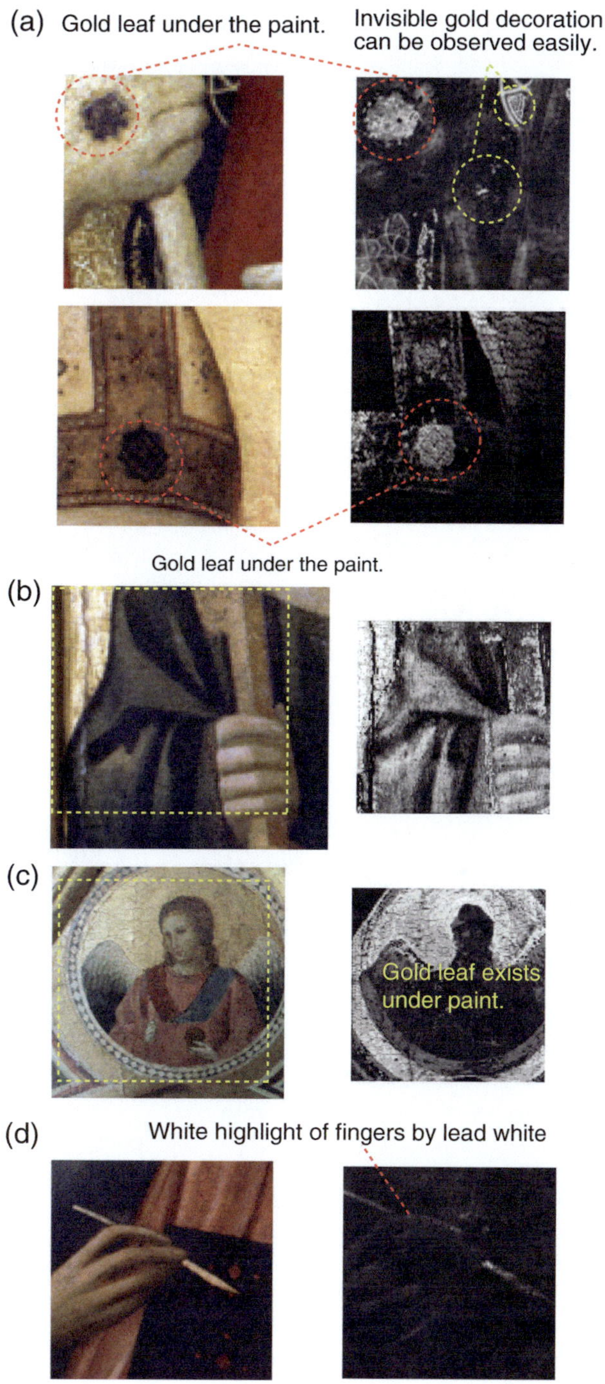

Fig. 6.18 THz pulsed TDI of parts of the panels of the saints and an angel from Polittico di Badia. (**a**) San Nicola, (**b**) Angel from the panel of San Pietro, (**c**) San Pietro, and (**d**) San Giovanni

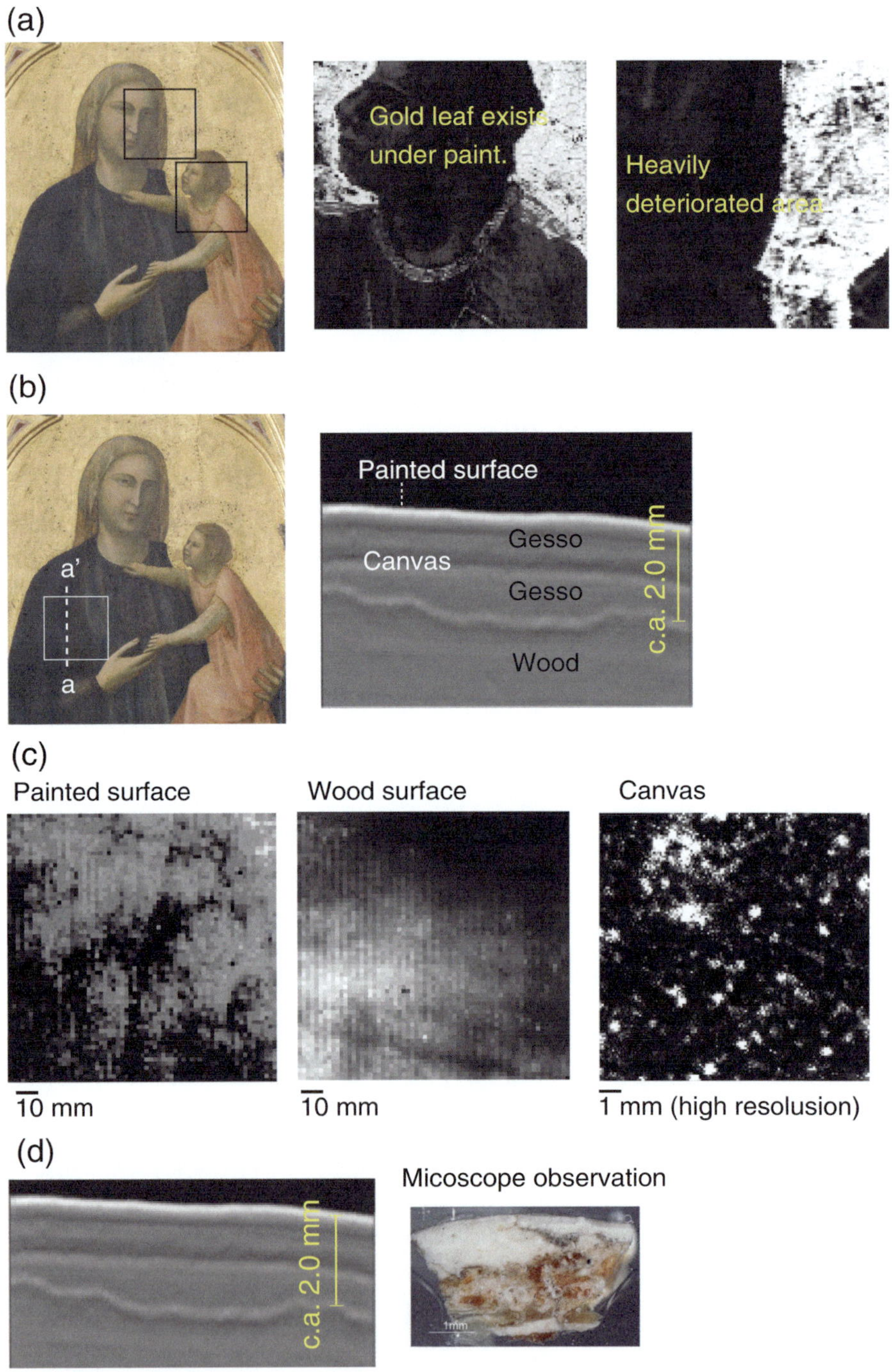

Fig. 6.19 THz pulsed TDI of a part from the panel of the Madonna with the Child. (**a**) Area image near the surface, (**b**) cross section along the *broken line* a-a′, (**c**) THz pulsed TDI of the area indicated by the *white square* in part (**b**) at the painted surface (*left*), the wood panel

reflection image of the surface is affected by the pigments, and the tool marks on the surface of the wood panel are visible. The high-resolution area image at the position of the cloth shows a regular mesh pattern, which supports the conservator's description of the process. The nondestructive cross-section image that was obtained by THz pulsed TDI is confirmed by comparison with the cross-section image of a sample taken from the fringe of the artwork by microscope observation, as shown in Fig. 6.19d.

A large crack was present in the panel of San Benedetto (Fig. 6.20). The edge of the crack where the wood support still existed appeared in the cross-section along the red line a-a′, as shown in Fig. 6.20a. The difference in rank is recognised at the edge of the crack. This figure also showed that the internal structure cannot be seen under the gold leaf. In areas where the metal leaf or gilding has separated slightly from the underlying layer or has been subject to surface deterioration, abrasion or friction, the reflected signal decreases. Therefore element analyses should be performed by XRF or other established methods to distinguish between the deteriorated gold and the yellow pigments.

Figure 6.20b shows another area in the panel of San Benedetto. The crack continued from the edge shown in Fig. 6.20a, and the cross-section along the red line at b-b′ clearly shows that the original wood support was missing. This indicates that a new wood panel had been inserted to compensate for the damage in the previous conservation effort, and that no layered structure was formed using cloth. The cross-section image revealed that this treatment actually caused a new problem, producing tension across the crack, which resulted in more severe deformation. After conservation treatment, the crack was not visible when looking at the artwork from the audience at a certain distance.

Another masterpiece composed of tempera on wood was also examined by THz pulsed TDI. The "Trittico di San Giovenale" (1422; central panel: 108×65 cm; each of the lateral panels: 88×44 cm) was painted by Masaccio, and is part of the permanent collection of the Masaccio Museum at the Pieve of San Pietro in Cascia di Reggello; the painting was also observed by THz pulsed TDI. Photographs taken during the conservation process and the measurement setup are shown in Fig. 6.21 [27].

The THz pulsed TDI observation area of the central panel of the Trittico is shown as a white square in Fig. 6.22a. The THz reflection image (Fig. 6.22b) shows the precise gold decoration of the mantle because of its high reflectivity level, while it is not clearly visible in Fig. 6.22a. While the differences between the pigments are insignificant in false colours that are determined simply by reflection level, some parts of the face and neck of the Madonna were more reflective than the other areas. This suggests that these highlights were painted with lead white, which has the highest reflection level among the white pigments. The cross-section image along

Fig. 6.19 (continued) surface (*centre*), and part of the canvas/cloth (*right*) in high resolution, (**d**) comparison between the cross-section images that were obtained by microscopy and by THz pulsed TDI

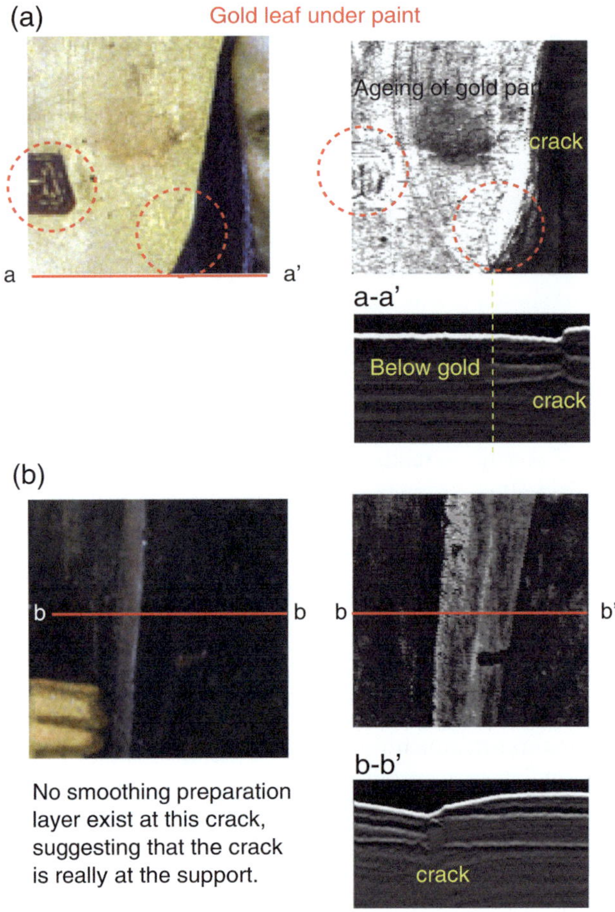

Fig. 6.20 THz pulsed TDI of the panel of San Benedetto. (**a**) Area image and cross-section near the face, (**b**) area image and cross-section near the hand, where a large crack exists

the line a-a′ indicated in Fig. 6.22a is shown in Fig. 6.22c. Some separations are recognised near the surface. These separations can be recognised acoustically by the conservator by simply touching the surface. Another significant feature is that the cloth layer is missing in some areas and the gesso was thus painted directly. However, the gesso layer is homogeneous at all points along the line a-a′, even at the edge of the canvas/cloth. This suggests that the gesso was applied at the same time in the areas with and without the canvas/cloth. When the area information at the cloth position was extracted from Fig. 6.22d, an unusual shape appeared in the image at a position that corresponded to the area without the cloth. According to the conservators, some additional pieces of wood were inserted, either in the original

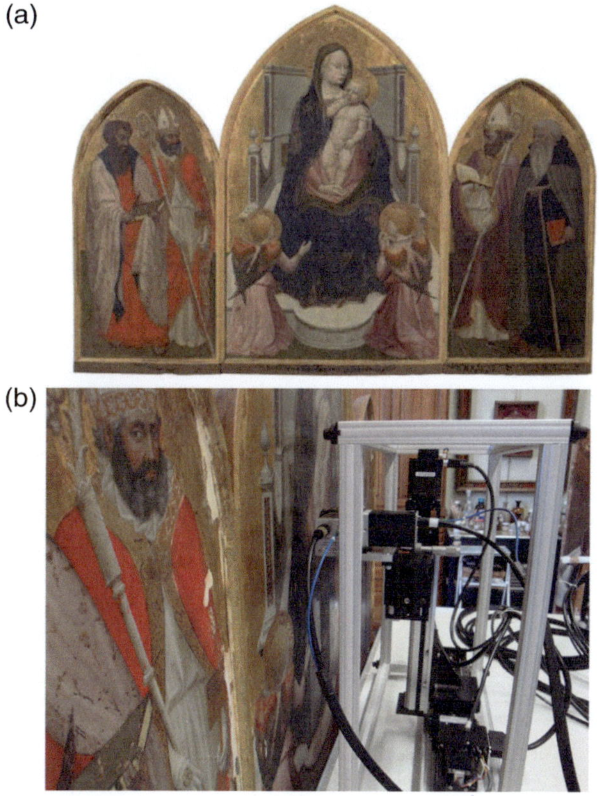

Fig. 6.21 THz pulsed TDI of "Trittico di San Giovenale" by Masaccio. (**a**) Photograph of the artwork, (**b**) measurement setup

artwork or as part of the previous conservation process, and thus it was concluded that the shape is formed by the presence of one of these wood pieces.

Figure 6.23a shows the THz pulsed TDI observation area in the panel of San Bartolomeo. As shown in the THz reflection image (Fig. 6.23b), the golden yellow coloured parts of the decorations of the mantles of the saint have little reflection. Because a few spots still present high reflection characteristics, it is considered that the area was originally covered by gold leaf which is almost entirely lost, and they were then painted with a yellow pigment. Unlike the gloves of San Pietro of "Polittico di Badia", which were shown in Fig. 6.18a, no gold leaf is hidden under the paint. When the book part was considered, the green colours of the book and those of the book latch show significant differences in their reflections. It suggests that both copper-based and green earth pigments were used in the original work or during the previous conservation process. The decoration at the centre of the book cover was lost and appeared as a dark spot in the THz reflection image. Cross-section images taken along the lines marked a-a′ and b-b′ in Fig. 6.23a are shown in Fig. 6.23c. This part of the painting had deteriorated severely, and was

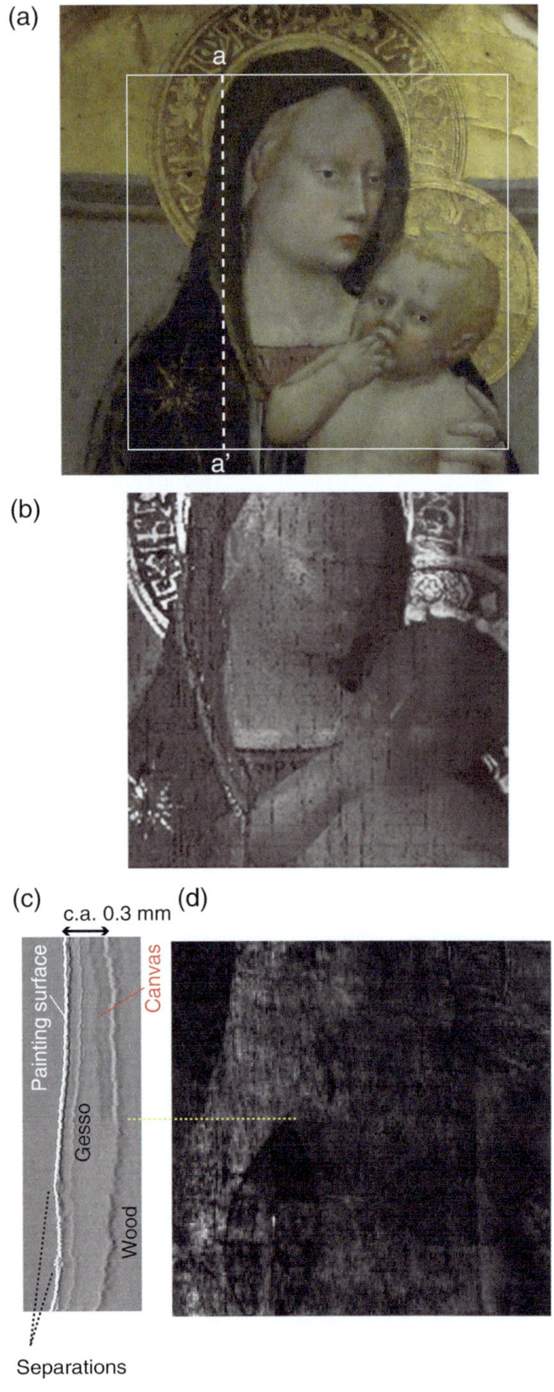

Fig. 6.22 THz pulsed TDI of the central panel. (**a**) Observation area, (**b**) THz reflection image, (**c**) cross-section image along the line *a-a*′ that was indicated in part (**a**), (**d**) Area image at the depth of the cloth layer

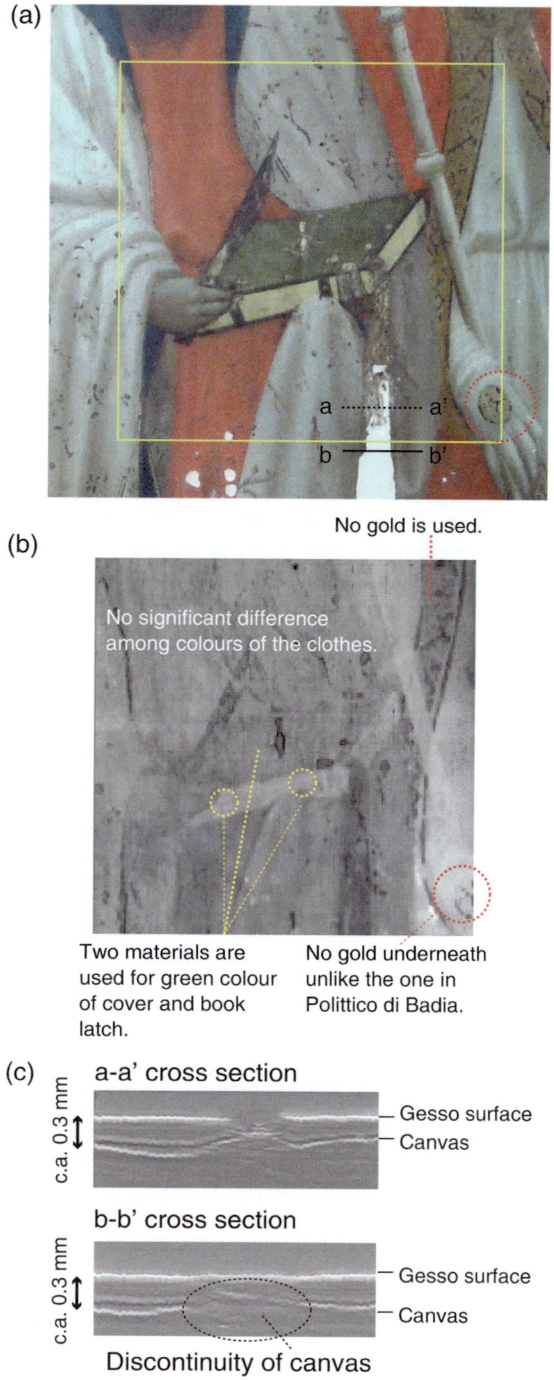

Fig. 6.23 THz pulsed TDI of the panel of San Bartolomeo. (**a**) Observation area, (**b**) THz reflection image, (**c**) cross-section images along the lines marked a-a′ and b-b′ in part (**a**)

under conservation, having been filled with gesso. The cloth layer remained along the a-a′ line, but the cloth was missing along the b-b′ line. In fact, there were many areas in this artwork where the cloth was missing.

The THz pulsed TDI of another section of the panel of San Bartolomeo is shown in Fig. 6.24. The high reflection levels of the background indicate that the gold leaf in this area is well preserved. The gold leaf also remains under the paint of the baton, and it appears as if the baton was almost painted on top of the gold leaf. In contrast, the decorations on the mantle and the mitra were painted with yellow pigments judging by the low reflection from those areas, although they should have been decorated with gold leaf. The frame area shows almost no reflection because the position was out of focus.

Similar features also appeared around the batons in the panel of San Giovenale and San Giovanni, as shown in Fig. 6.25b. Gold leaf was still present in the background, and could clearly be observed even under the paint of the baton. In contrast, the decorations on the mantle and the mitra were painted with yellow pigments. The dark spots that appeared in the gold area were caused by surface scattering at spots on the gold as "incisione", indicating that the background gold in the artwork is basically well preserved. This evidence suggested that the replacement of the gold with the yellow pigments in the clothes and hats of the saints was possibly done in previous conservations.

In the case of oil panel paintings, the preparation layer is much thinner than that in tempera painting. It is therefore expected that THz pulsed TDI would not be able to distinguish between the painting and the preparation layers. Figure 6.26 shows an oil painting on wood, "Ritratto di Ragazza", by Rosso Fiorentino (1514), which is part of the permanent collection at the Uffizi Gallery. A typical example of a cross-section image shown in Fig. 6.26b shows that the structures of the painting and preparation layers are uniform, and no particular deterioration, such as separation, is observed. It is too thin to be distinguished layer-by-layer by using THz pulsed TDI, the area images below the surface can be observed by extracting reflection pulses. Figure 6.26c, d show the area image below the varnish and the wood surface, respectively. While it was difficult to identify the pigments from the THz image shown in Fig. 6.26c, the high reflection area in the background suggested that it was painted with lead white. The tree rings are clearly observed at the surface of the wood support as shown in Fig. 6.26d. The black spots that can be seen in this figure are most likely made from insects, because the spots appear continuously along the depth direction of the wood support [28]. Successive works have demonstrated the high potential of THz pulsed TDI for use as an analytical tool for panel paintings [29, 30].

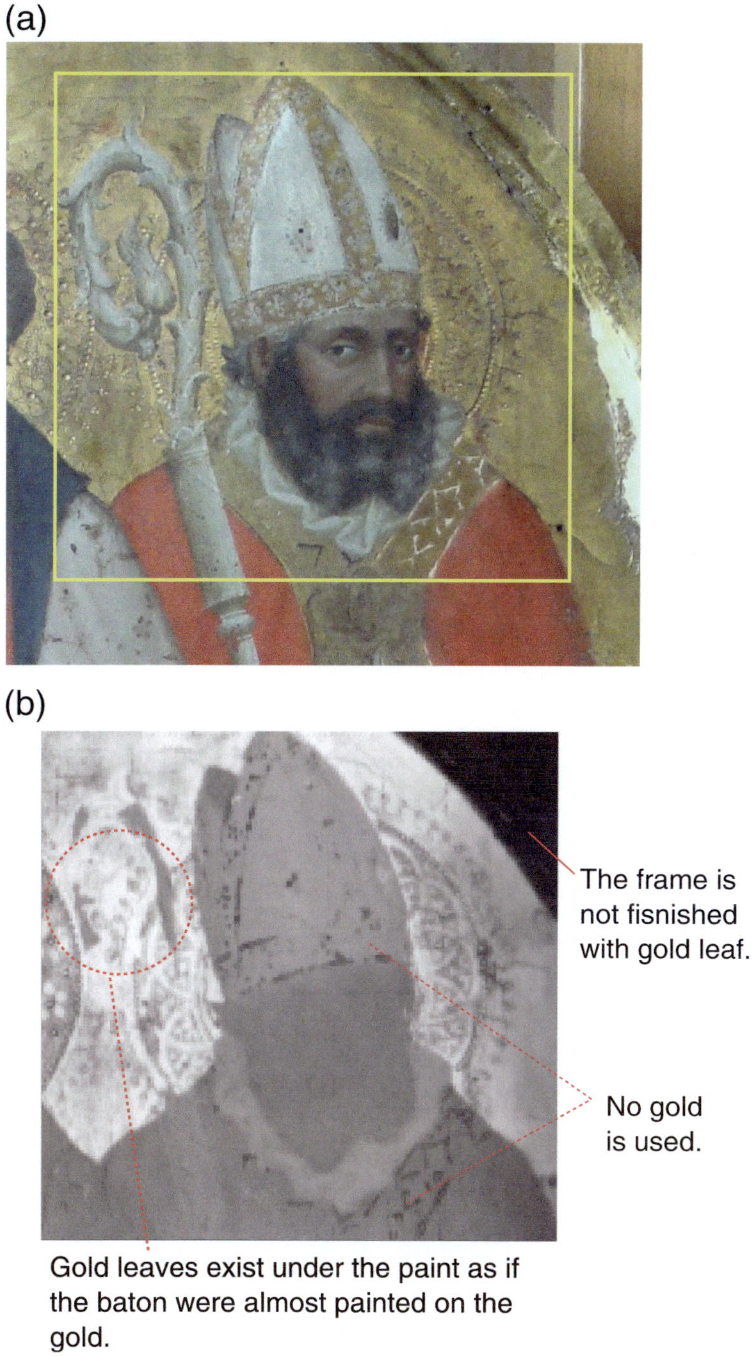

Fig. 6.24 THz pulsed TDI of the panel of San Bartolomeo. (**a**) Observation area, (**b**) THz reflection image

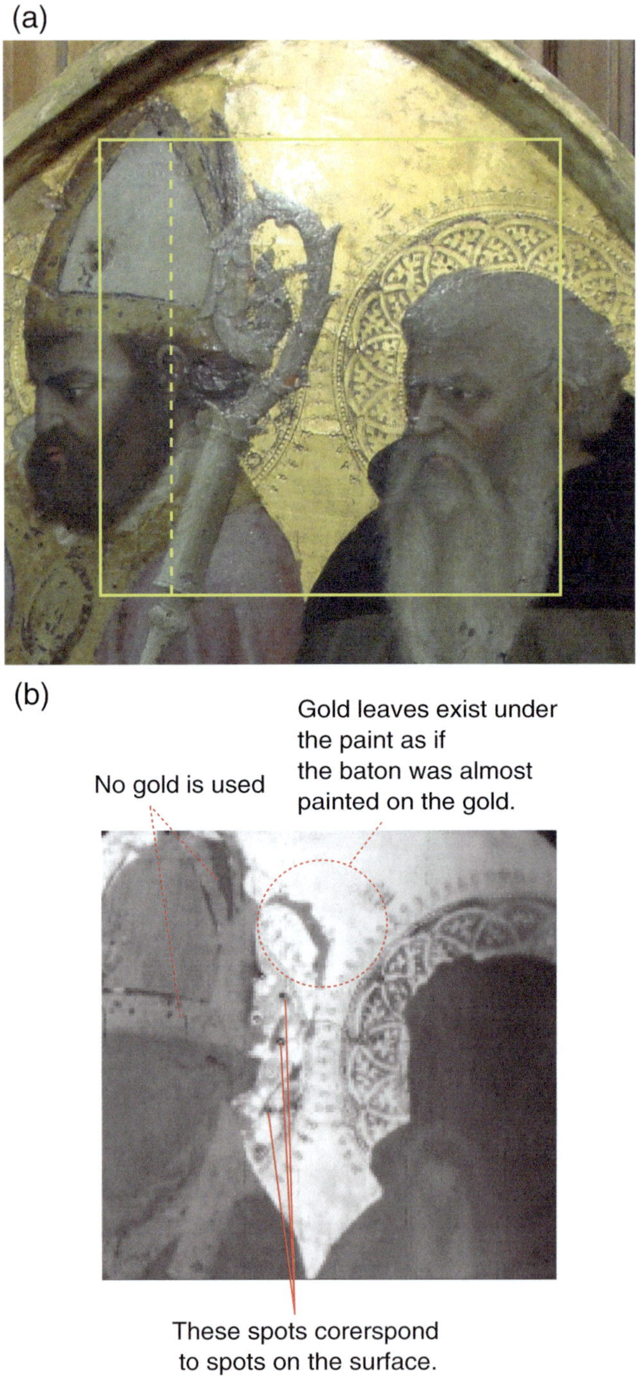

Fig. 6.25 THz pulsed TDI of the panel of San Giovenale and San Giovanni. (**a**) Observation area, (**b**) THz reflection image

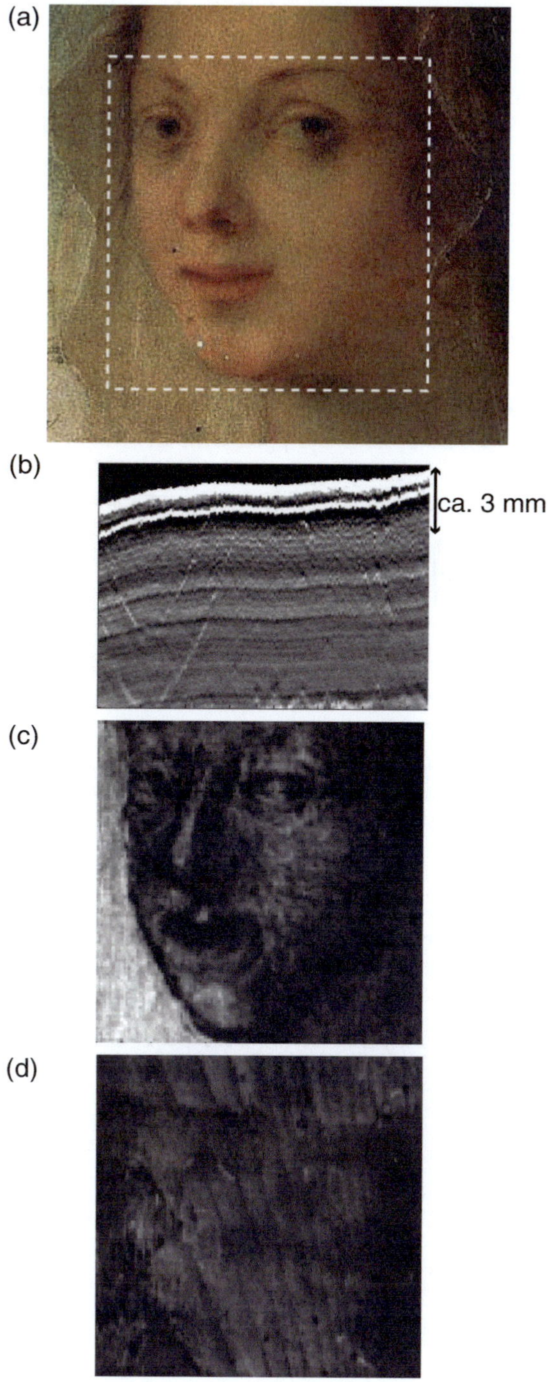

Fig. 6.26 THz pulsed TDI of "Ritratto di Ragazza" by Rosso Fiorentino. (**a**) Observation area, (**b**) typical cross-section of the wood panel painting, (**c**) area image below varnish, (**d**) area image of the wood surface

6.4 Japanese Panel Paintings

Paintings on wood are also found in Japan, often as decorative walls or pillars in the wooden architecture of buildings such as temples. The preparation layers for this type of decoration are formed with lime white, which is made from seashells. The principal substance in lime white is calcium carbonate. The most common deterioration process in these paintings is the separation of the painting layer and the lime white preparation layer from the surface of the wood. Similar to western panel paintings, cloth is often applied to the wood to prevent the generation of cracks on the surface, which are caused by deformation of the wood due to changes in humidity and temperature.

As introduced in Fig. 4.15 in previous chapter, the curvature of the support affects the THz area image in the case of pillar observation when using a THz pulsed TDI system equipped with an ordinary X-Y stage to control the sensor unit position, as shown in Fig. 6.27. The pillar under observation is located in the inner sanctum of a three-storied pagoda of Kiyomizu-dera, which was firstly built in the eighth century in Kyoto, Japan. The temple and pagodas are registered as part of the Historic Monuments of Ancient Kyoto UNESCO World Heritage site.

The THz reflection image of part of the painted surface of the pillar is compared with the photograph, as shown in Fig. 6.28. The colours of the painting remain in good condition, and in previous conservation treatments, the separated parts were all stuck together with natural glue. However, the severe environmental conditions in and around the pagoda can cause separations. As shown in Fig. 6.28b, some high reflection areas appeared, as indicated by the yellow oval. It is due to the air in the separation and superposition of the visible and THz reflection images (Fig. 6.28c) reveals the areas that require additional treatment with glue.

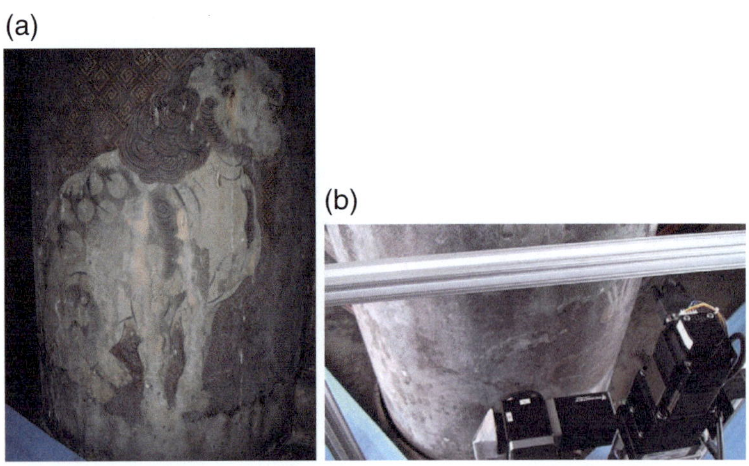

Fig. 6.27 Pillar of the inner sanctum of a three-storied pagoda of Kiyomizu-dera, Kyoto. (**a**) Photograph of the part under observation, (**b**) measurement setup

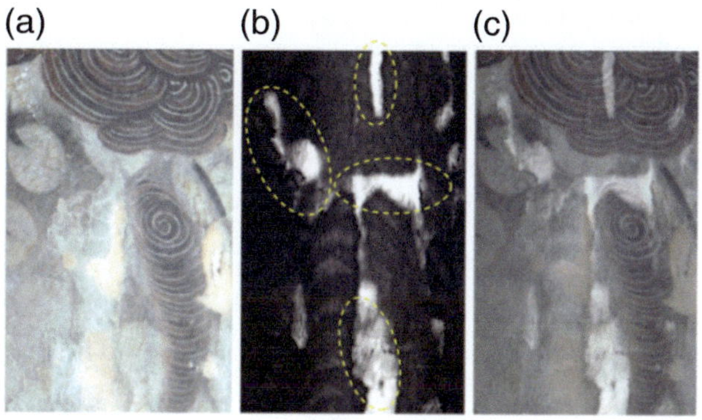

Fig. 6.28 THz pulsed TDI of the pillar under investigation. (**a**) Observation area, (**b**) THz reflection image, (**c**) superposition of the THz image on the photograph

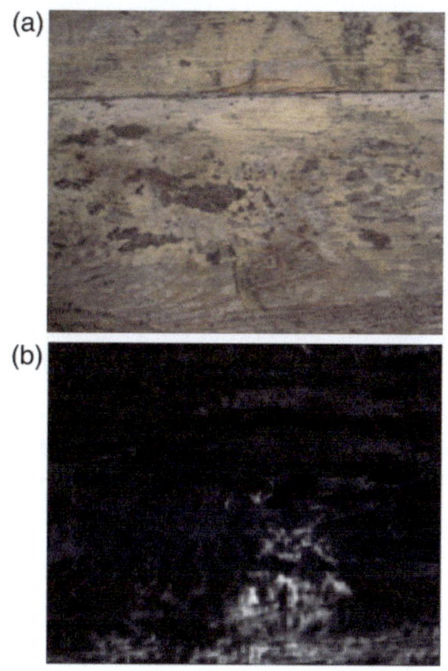

Fig. 6.29 THz pulsed TDI of a horizontal piece of timber in the inner sanctum of a three-storied Koyasu pagoda. (**a**) Observation area, (**b**) THz reflection image

The reflection levels from the surface can not only be used to detect separation, but can also indicate information about the pigments on the surface in certain cases. Figure 6.29 shows an image from the observation of a horizontal piece of timber in the inner sanctum of a three-storied Koyasu pagoda. A high reflection area appeared in the bottom central area of the THz reflection image as shown in Fig. 6.29b. This indicates that some paint containing mineral pigments remained in that area, although it is not visible in the photograph (Fig. 6.29a). According to historians, the pagoda was initially constructed with a vivid colour scheme, and was then painted red and later black. The latest conservation process (2009–2012) returned the colour scheme to the original vivid colours based on historical records.

The next example is a panel painting on a wooden door with a peacock motif, located in the outdoor terrace of the Tanzan Shrine, which was established in 678 CE, and has been renovated several times during its history. The main shrine

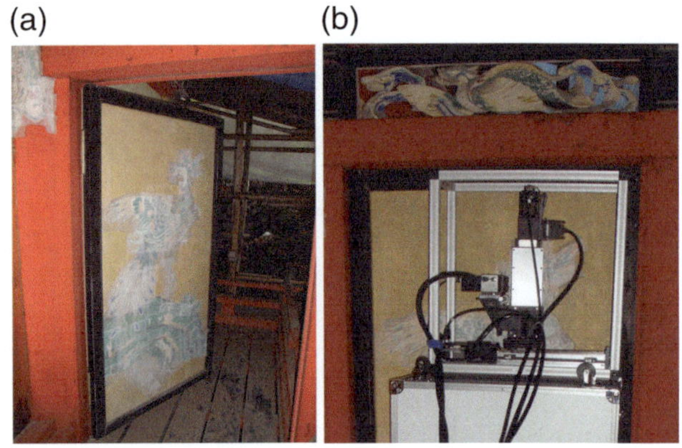

Fig. 6.30 THz pulsed TDI of painted wooden door in the main shrine of Tanzan Shrine. (**a**) Photograph of part of the outdoor terrace, (**b**) measurement unit

was under conservation from 2010 to 2012. A thin paper layer protected the painting area, except for the gold leaf background area during the conservation of the building. The measurement unit shown in Fig. 6.30b actually required a heating system, because the outside temperature was too low for the internal laser to begin oscillation.

The section of the wings of a peacock (Fig. 6.31a), was also observed by THz pulsed TDI. The THz reflection image and the superposition with the visible image are shown in Fig. 6.31b, c, respectively. The protective paper is almost transparent in the THz region, similar to the piece of thin cloth that was introduced in Fig. 6.1. The strong reflection from the area where the colour pigments are missing proves that the area was painted on the lime-white preparation layer on gold-finished wood.

The white plumage of the wings was painted using lime white on the gold-finished wood, and the same pigment was also used to form a preparation layer for other deep colours as described in Fig. 6.31. The vivid green and blue colours were only achieved by using coarse pigments of azurite and malachite on lime white. Direct painting on gold leaf requires extremely fine particles, and thus lime white is extremely well-suited for this purpose. The spots that exist in the blue area were possibly made from additional gold leaf, and black lines drawn were painted by carbon black, because it is not visible in THz pulsed TDI. The technique of painting on gold is also discussed in Sect. 6.5.

THz pulsed TDI of the head part of the peacock is shown in Fig. 6.32, which is arranged in a similar manner to the images in Fig. 6.31. The pupil of the peacock's eye was probably finished with additional gold leaf, judged by its high reflection. Some of the red coloured parts have slightly higher reflective levels than the green and blue pigments, and thus the red pigment is expected to be a mineral pigment such as cinnabar. As per the example of Fig. 6.31, the black lines drawn using carbon black and the protection paper are not visible in the THz reflection image.

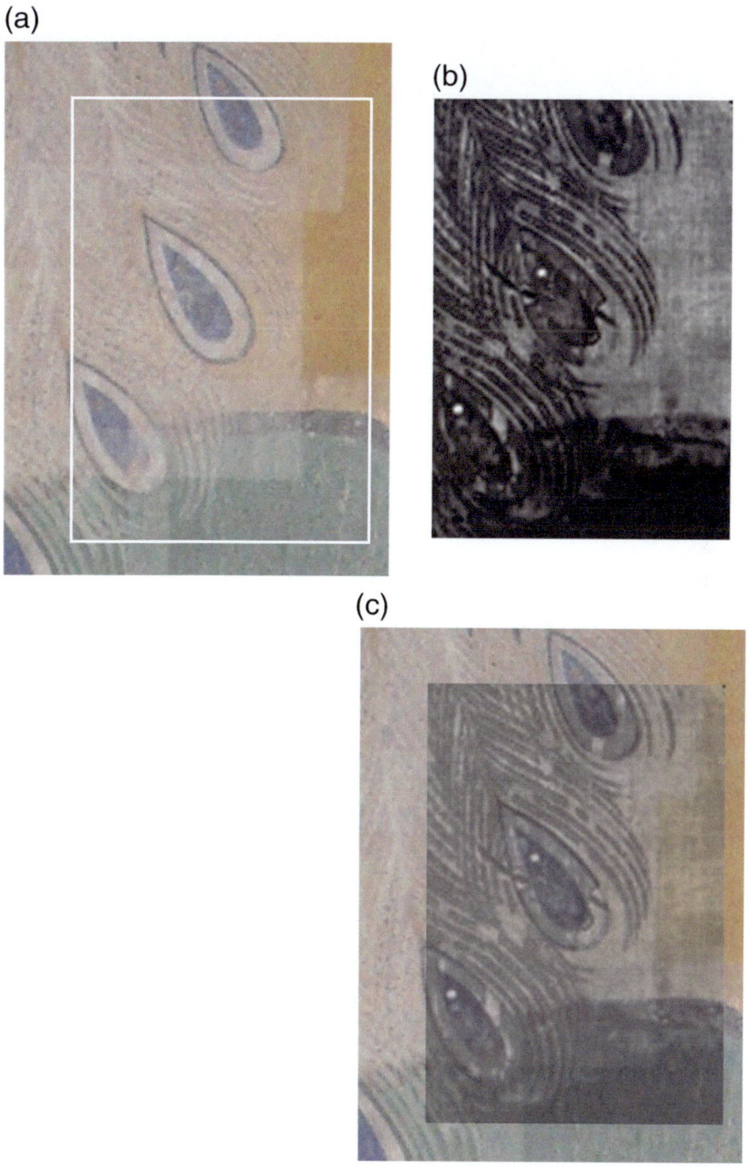

Fig. 6.31 THz pulsed TDI of a painted wooden door in the Tanzan Shrine. (**a**) Observation area around the wings of a peacock, (**b**) THz reflection image, (**c**) superposition of THz image on photograph

In this example, gold leaf covered the entire surface of the door on the both sides, and thus the internal structure could not be observed. This is a common disadvantage when using electromagnetic waves to analyse metallic objects.

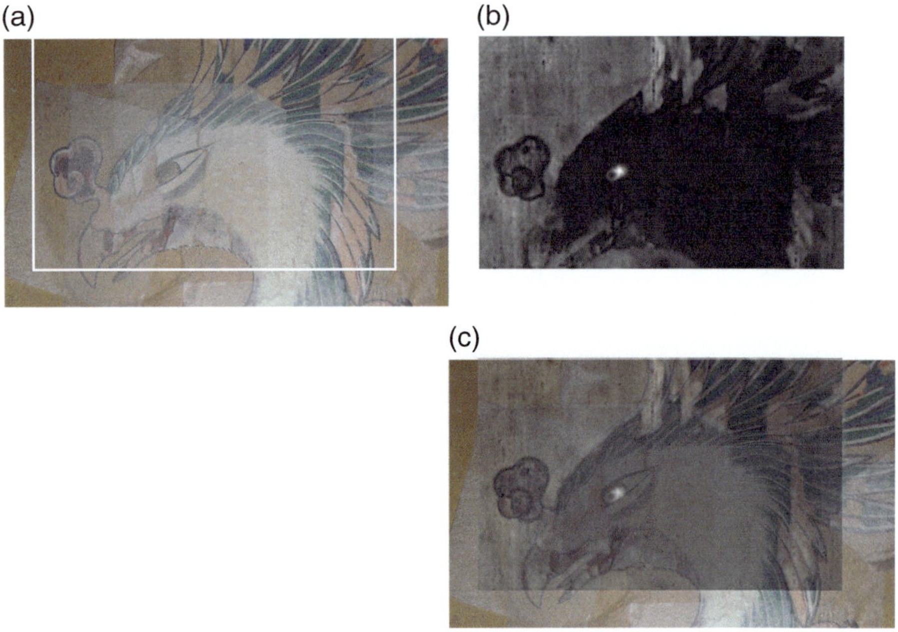

Fig. 6.32 THz pulsed TDI of a painted wooden door in Tanzan Shrine. (**a**) Observation area around the peacock's head, (**b**) THz reflection image, (**c**) superposition of THz image on photograph

6.5 Japanese Panel Screens

Japanese panel screens and sliding doors are generally composed of layers of paper on a latticed wooden mount. The paper layers are glued using starch. In general, the layers are made for three specific purposes. The bottom layer that is fixed to the wooden mount is used to form the overall shape, and thus relatively thick and strong but rather rough paper is used. In the middle section, thin and soft paper layers are partially glued to each other with air gaps to form an internal tension absorber. This layer works to ease the tension of the surface painting paper when the wooden mount deforms along with changes in temperature and humidity. The top painting paper is made from a strong but smooth paper, and is partially glued to the middle layer. The preparation layer of lime white is applied to the painting paper before it is painted with pigments and animal glue. The uneven relief-like surface is made by lime white as a part of preparation layer, and it is then covered by gold leaf [31]. Most sixteenth to nineteenth century Japanese masterpieces have such features, which give a particular effect when the direction of the ambient light changes.

Full-scale conservation of these artworks involves the replacement of all supporting parts, from the middle "absorber" part made from multi-layered paper to the latticed wooden mount. Because the top painting paper is partially glued, it

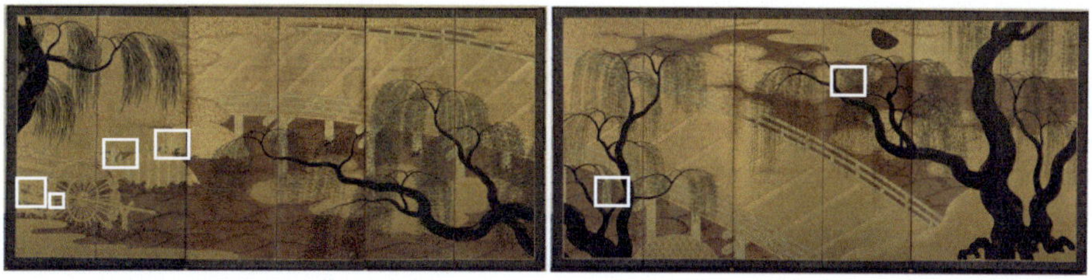

Fig. 6.33 THz pulsed TDI observation area of a Japanese panel screen, "Willow Bridge and Water Wheel", dated seventeenth to eighteenth century, from the collection of the Tokyo National Museum

can be removed using a wooden paper knife without high risk. Before the top painting paper is removed, its surface is protected by applying thin pieces of paper using glue made from a marine alga. This treatment also works to remove stains from the surface.

While the condition of the painting itself can simply be examined by eye, the internal preparation layer is not visible. The THz pulsed TDI was applied to observe internal structures under gold leaf cannot be observed using electromagnetic waves, but a panel screen that only has gold leaf on the surface side can be observed from the rear side. Because Japanese panel screens commonly have a thickness of approximately 20 mm, THz pulses can pass through all parts of the screen except for the latticed wooden mount.

Figure 6.33 shows the first panel screen observed by THz pulsed TDI, the "Willow Bridge and Water Wheel" (unknown artist, c. seventeenth century), from the collection of the Tokyo National Museum [32]. As mentioned above, panel screens can be observed from both sides, and an example of this is shown in Fig. 6.34. Red drawing lines on the barrel, which appeared in the photograph shown in Fig. 6.34a, are not visible in the THz reflection images shown in Fig. 6.34b, c. The reflection from gold leaf becomes highest in the observation area. This suggests that the lines are painted on gold using one of the mineral pigments or dyes that are relatively transparent in the THz region. The brown areas shown in both THz images have no gold leaf on the surface or under the paint. The edges of the relief-like preparation using lime white generate surface scattering, which resulted in a reduction of the THz reflection level. Thus, even when the surface is completely covered with gold, lines appear in the THz reflection images. The latticed wooden support disturbs the THz pulse transmission, and appears like a shadow, as shown in Fig. 6.34c.

The specific internal structures of the three parts appeared clearly, as shown in Fig. 6.35. Similar to Fig. 6.34c, the edges of the lime-white decorative reliefs appear as lines in the THz reflection image in Fig. 6.35b, e, and this effect is confirmed by the cross-section images of the front layers of the screen that are shown in Fig. 6.35c, f. The total thickness of the three parts of the multi-layered paper is approximately 3 mm. This value is estimated by assuming that the velocity

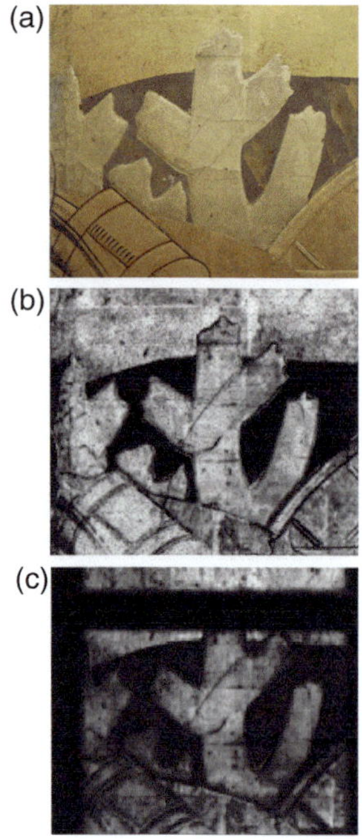

Fig. 6.34 THz TDI of a section of "Willow Bridge and Water Wheel". (**a**) Observation area, (**b**) THz reflection imaging from front side, (**c**) THz reflection imaging from rear side

of the THz pulse is the same as that in air. As shown in Fig. 6.35c, in the middle section shown as part B, detached pieces of paper are recognised because they are partially glued to produce a bellows-like form. When the THz pulse reaches the surface that is covered with gold, multiple reflections often occur inside the preparation layer, and these signals appear as ghost images, as shown in Fig. 6.35c, f. When the lime-white reliefs are in good condition, they appear clearly as independent preparation layers. When the lime-white reliefs deteriorate, these parts produce irregular reflections due to internal scattering within porous structure, as observed in Fig. 6.35f.

While internal structure observation is the most important task for THz pulsed TDI in the examination of panel screens, measurements from the front also provide useful information for practical conservation planning. Figure 6.36b shows a THz reflection image where most of the area is covered with gold, as shown in Fig. 6.36a. Similar to the previous examples, lines appear in the gold area at the edges of the lime-white reliefs. The areas marked in red dotted ovals in Fig. 6.36a appear to be gold to match their surroundings. However, the lack of reflection from these areas, as shown in Fig. 6.36b, revealed that these areas are not covered with gold leaf. They were actually painted with yellow paint, which most likely occurred as part of a previous conservation treatment. There is also no reflection in the area of the water wave, which appears as a dark image, and thus there is no gold leaf underneath that section either. However, some water splashes, indicated by the

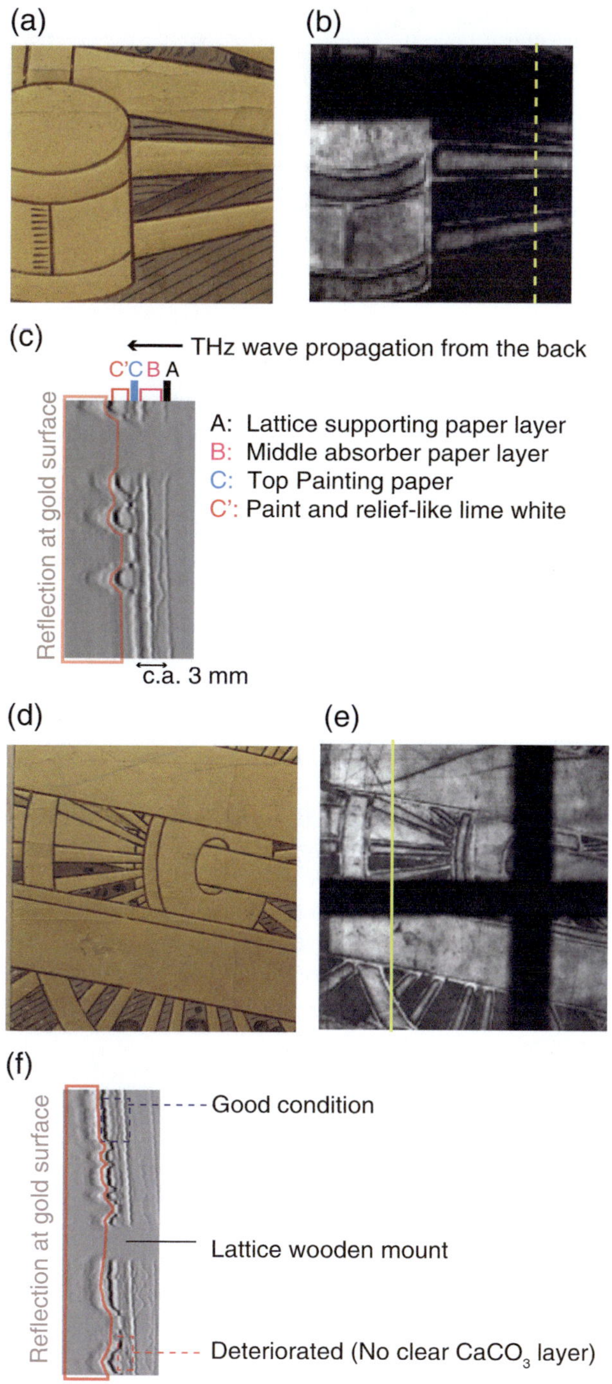

Fig. 6.35 Internal structure observation by THz pulsed TDI from rear side of panel screen. (**a**) Observation area with water basket, (**b**) THz reflection image from rear, (**c**) cross-section image along the *yellow broken line* shown in part (**b**), (**d**) observation area of part of the wheel, (**e**) THz reflection image from rear, (**f**) cross-section image along the *yellow solid line*

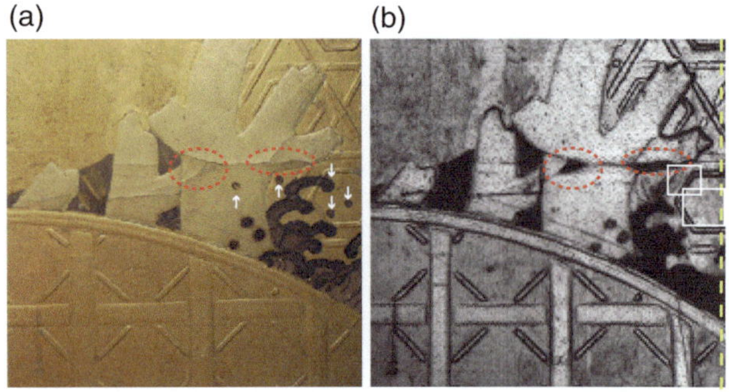

Fig. 6.36 THz pulsed TDI of area of panel screen viewed from the front. (**a**) Observation area, (**b**) THz reflection image

white arrows, did not appear as dark spots in the THz reflection image. These areas could be composed of carbon black painted on gold.

Figure 6.37 clearly shows the different techniques applied when using gold leaf. The background of the willow trunk appears to be almost covered with gold leaf, as shown in Fig. 6.37a. However, the THz reflection image in Fig. 6.37b reveals that small square pieces of gold leaf have been glued in place. There is no gold underneath the trunk of the willow or the wave, judging by the lack of reflection. The branches are not recognised in the THz reflection image, while the leaf appears as shadow spots on the gold parts. It was therefore suggested that the branch was painted using carbon black and probably included earth pigments that have high transmittance in the THz frequency region, while the leaf is painted using relatively coarse malachite pigments that cause scattering, which results in a reduction of the reflection level.

The bridge girder (Fig. 6.37c), however, is covered with gold leaf, and like the leaf areas in Fig. 6.37b, shadow-like images appear in the corresponding THz reflection image, as shown in Fig. 6.37d. The advantage of THz pulsed TDI is that an area image can be obtained by extracting the THz pulse in the time domain, as described in previous chapters. The signals that are detected before the largest signal from the gold surface signify the reflections from the leaf, and the map of these reflections shown in Fig. 6.37e shows good agreement with the visible images.

When the area is not covered with gold leaf, the internal structure can be obtained via measurements from the surface. By extracting the signals from the latticed wood mount, the edge of the supporting paper can be observed, as shown in Fig. 6.37f.

"Irises at Yatsuhashi" by Ogata Korin is one of the masterpieces of his particular school. The history of and artistic techniques used in the piece have been widely discussed in various books and technical papers, and the artwork itself was recently examined [33]. This artwork is part of the collection of the Metropolitan Museum of Art in New York, and was examined by THz pulsed TDI before an exhibition in Tokyo in 2011. In addition to examination of the latticed wood frame and the

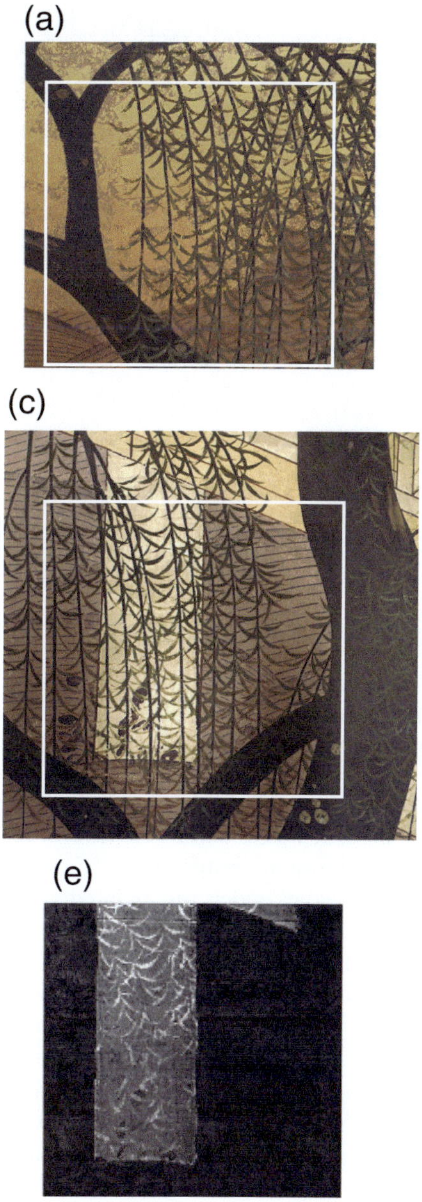

Fig. 6.37 THz pulsed TDI of an area of the panel screen from the front. (**a**) Photograph of willow trunk area, (**b**) THz reflection image, (**c**) photograph of bridge girder area, (**d**) THz reflection image, (**e**) THz reflection above the gold leaf, (**f**) THz reflection from the depth of the latticed wooden mount

(a)
(b)

Fig. 6.38 "Irises at Yatsuhashi" by Ogata Korin. (**a**) Digital image using high resolution photographs, and shown in the diagonal direction, (**b**) measurement setup

supporting paper layers, including the detection of insect holes, clarification of the technique used for the gold leaf in this artwork was a major objective of the study.

In the case of another masterpiece with the same motif, "Irises" from the Nezu Museum in Tokyo, no gold leaf was detected under the painted area during conservation work in the period from 2003 to 2005 [34]. In contrast, high-resolution images of "Irises at Yatsuhashi" indicated that there could be gold leaf under the paint. Because the detection of gold leaf under paint is one of the main advantages of THz pulsed TDI, as described in previous chapters, the artwork was examined by this method, as shown in Fig. 6.38.

In this figure, a diagonal photograph is used to enhance the artistic value. The composition and the distribution of the irises were designed well to fit the panel screen shape. When the painting is displayed as a flat shape, as it would be in a common catalogue, the perspective and the spatial effects are reduced.

The areas for internal structure observation are chosen around the parts that have visible defects, as shown in Fig. 6.39a, d. The defects at the part of the bridge shown in Fig. 6.39a are expected to be limited near the surface, and the THz pulsed TDI measurements were performed from the rear with the focus at the surface. Figure 6.39b shows the THz reflection image of the area indicated by the red dotted square in Fig. 6.39a, and the cross-section of the front half of the panel along the line a-a′. The figure confirmed that the holes in the visible and THz images along

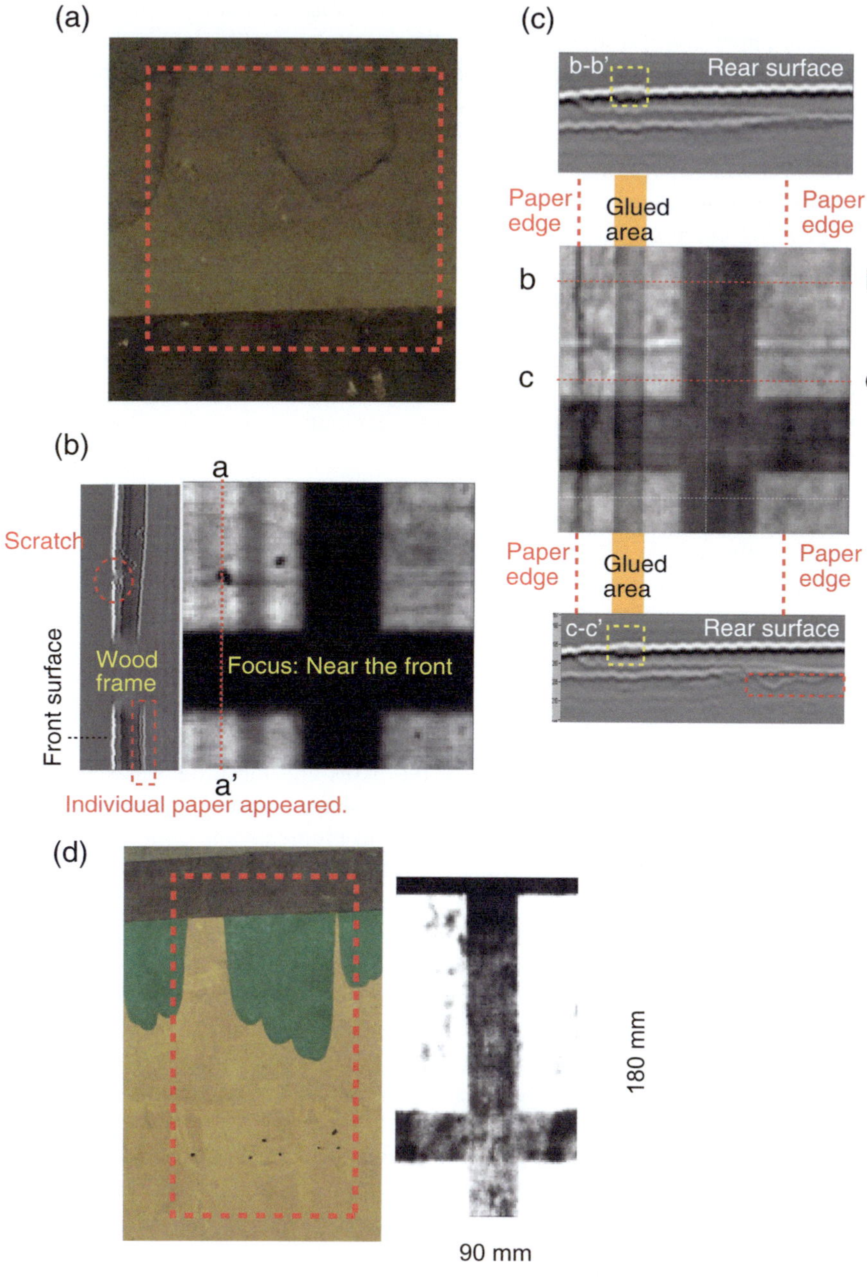

Fig. 6.39 THz pulsed TDI of part of the panel from the rear. (**a**) Photograph of bridge area under observation, (**b**) THz reflection image focused at the surface with the cross-section image of the front half side, (**c**) THz image of bridge area focused at the wood mount and the cross-section images of the rear side half, (**d**) photograph and THz image of area where insect holes were found

the line a-a′ only exist at the surface, and the holes did not reach the middle "absorber" paper layer, so the defects are thus scratches in the gold leaf.

As shown in Fig. 6.39, the latticed wooden mount interfered with the observation from behind the mount. The condition of the wood can be estimated by setting the focus at the surface of the wood frame from the rear side, although the applicable depth is limited to approximately 5 mm or less. The propagation of THz waves in wood has been discussed in detail in previous publications, as mentioned in the panel painting section (Sect. 6.3). It may be possible to observe the structures through the latticed wooden mount by changing the incident wave direction, this may not be required in examinations before conservation planning for panel screens in practice. Because the wooden mount itself will be removed for replacement.

To observe the internal paper layer structure, the focus was set at the back side of the structure. The cross-section of the rear half of the panel was observed, as shown in Fig. 6.39c. For example, two paper edges, where one is near the surface and the other is in the middle, are clearly shown in the cross-section images and in the area image when the focus is set at the wood mount, which is indicated by the red broken line. The line of the paper edge indicated that the paper was cut in a straight line.

Additionally, a wide straight line also appeared in the area image. The edge of this line is visible at the rear surface, and it is the edge of the rear surface paper. Based on the cross-section image, the wide line was considered to be the margin of the glue.

Holes appeared in the area that was covered with gold leaf, as shown in Fig. 6.39d are made by insects which generally eat the wood, and move along the lattice. The holes are thought to be made when the insects develop from larvae into adults. In this figure, the green leaves of the irises and the brown bridge do not appear in the THz reflection image. This is because the painting paper was entirely covered with gold leaf before being painted with pigments and animal glue. THz waves that pass though the paper layers are reflected at the gold layer, and cannot reach the paint from the rear side.

As shown in the cross-section image, the reflection from the surface is extremely strong, even though the THz wave was emitted from the rear side. The reflection level does not change in the area of the bridge or in the area of the leaves, neither of which are visibly gold. This behaviour proves that the surface of the artwork is entirely covered with gold leaf.

THz pulsed TDI from the surface (Fig. 6.40) confirmed that the gold leaf existed under the paint. Part of a flower and the bridge girder are barely visible in the THz image. The THz reflections from the painted area, such as those from the petals of the irises, show even higher reflection. This may be due to the protection of the gold surface by the paint, as in San Pietro's glove (Fig. 6.18). When we consider the cross-section, the entire panel screen is covered with gold, and thus the internal structure cannot be observed via measurements from the front, except at the pin-holes shown in the cross-section image of Fig. 6.40b.

Because the positions of the gold leaf areas are very clear in the time domain and the surface is flat, the thickness of the paint on the gold leaf can be estimated by extracting the peak that appears before the peak from the gold surface in the time

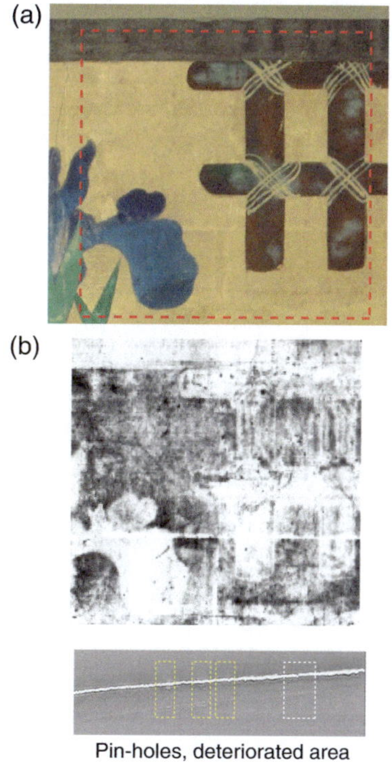

Fig. 6.40 THz pulsed TDI of part of the panel from the front. (**a**) Photograph of bridge girder and flower, (**b**) THz image of the area indicated in part (**a**) and the cross-section from the front

domain. Figure 6.41a shows the iris leaves with thick paint. First, the THz reflection image of the red broken area is obtained using the entire output signal. Figure 6.41b indicates only the flat, relatively uniform and high reflection area. The output signal from the paint, which should be detected before the larger reflection from the gold in the time domain, is too small to be recognised in Fig. 6.41b. However, the cross-section image that is placed above the area image in Fig. 6.41b clearly shows the paint layer on the surface that is covered with gold leaf. If the reflective index of the green paint (malachite) is assumed to be 1.7, then the thickness of the paint is approximately 0.6 mm. That thickness can be achieved by painting layer by layer while progressing from very fine pigments to coarse pigments. The reflection pulse from the paint layer appears prior to the large reflection from the gold surface, as shown in Fig. 6.41c. The reflection pulse at point A can be clearly seen to have two peaks, while a single peak from the gold leaf appeared at point B. The small negative peaks that appeared in the reflection pulse at point A are generated by the multiple reflections in the paint layer, and thus can be treated as ghost signals. The image of the paint layer itself can be extracted by using the pulse that appeared before the strong pulse from the gold surface in the time domain. The image obtained by this extraction method becomes the same as that of the visible iris leaves, as shown in Fig. 6.41d.

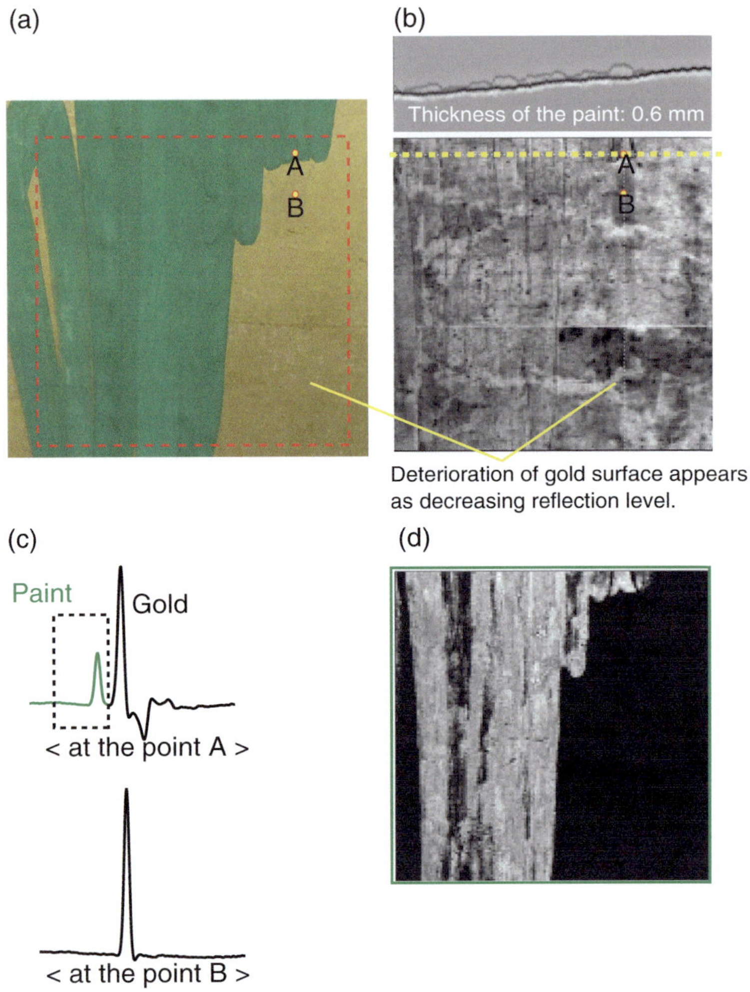

Fig. 6.41 THz reflection imaging of leaves with thick paint layers from the front. (**a**) Observation area; reflection pulses at points A and B are shown in part (**c**), (**b**) THz reflection image and cross-section along the *yellow dotted line*, (**c**) waveforms at points A and B in part (**a**) after deconvolution, (**d**) extracted image of paint layer

Although painting on gold leaf sounds like it would require high artistic skill levels, it can be done systematically to produce ready-made style artworks. Based on the information obtained by THz pulsed TDI and on the viewpoints of some conservators, the school of Korin established manufacturing techniques for their designed products to meet the high demand for these products. Reduction of the time required for cutting and adjustment of the gold leaf, depending on the design, could help to compensate for the cost of the gold leaf. However, the well-calculated design of the motif and the delicate expression of the centre of the flower with the warm colour pigments prove that the artwork is a masterpiece of Ogata Korin and his era.

The Cypress Tree by Kano Eitoku is one of the artist's masterpieces, and is registered as a national treasure by the Japanese government. The artwork was once a pair of sliding doors in a castle. Successive owners all loved this artwork, and modified it to fit their own castles. Finally, the work was converted into a transportable panel screen.

THz pulsed TDI was performed before the conservation process to observe the internal condition of the panel screen. It was anticipated that a repair treatment had been performed near the catch of the sliding door when the original form was changed into the panel screen. THz pulsed TDI revealed many rectangular pieces of paper that appeared below the painting, and the layer structure was observed in the cross-section image. The mending paper map that was obtained by THz pulsed TDI was compared with a map recorded by a conservator during the repair treatment, and the two maps were perfectly matched. This work is described in detail in [35].

Areas for observation by THz pulsed TDI are indicated in Fig. 6.42a, where the white and black squares are those areas observed from the front and from the rear,

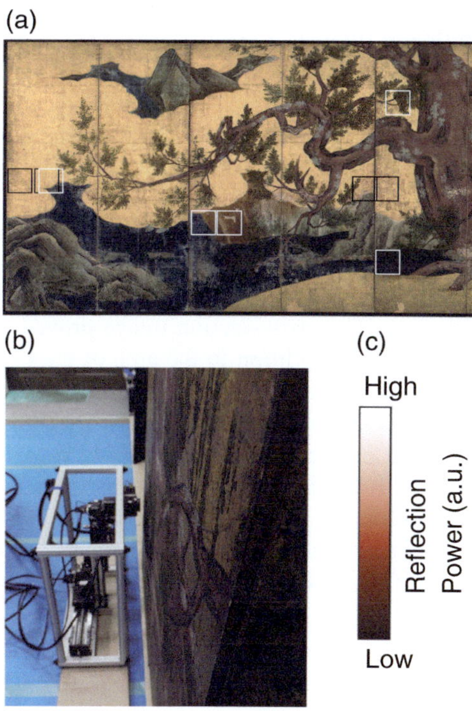

Fig. 6.42 "Cypress Tree" by Kano Eitoku. (**a**) Photograph before conservation process, showing observation areas, (**b**) measurement setup, (**c**) colour scale for THz images in this section

respectively. The instrument settings are shown in Fig. 6.42b. The THz reflection images, however, are not described using a simple grey scale; instead, a three-colour description (Fig. 6.42c) is provided to display the details more clearly. In this book, typical examples are shown to introduce previous conservation efforts that were observed by THz pulsed TDI.

THz pulsed TDI of the area shown in Fig. 6.43a is performed from both the front and rear sides. Figure 6.43b shows the THz reflection image from the rear side. This image indicates that the painted area contains no gold leaf underneath the paint, because the reflection level of the blue area is much lower than that of the gold area. The cross-section images along the line marked a-a′ show the typical internal structures of Japanese panel screens, and the papers are well adhered to the wooden mount. The structure near the surface looks more complex than the area beneath the gold leaf. Figure 6.43c shows the area image of the middle layer, which lies approximately 1.0 mm below the surface, and is indicated in the cross-section image by a blue broken line.

The complex layer structure near the surface was examined by THz pulsed TDI from the front side, although no information could be obtained from the area under the gold leaf. Figure 6.43d shows the THz reflection image approximately 0.5 mm below the surface, i.e., around the complex layers. A rectangular shape was clearly observed, and is expected to be a mending paper piece, along with the glued line of the support paper. The cross-section image shows the change in the number of paper layers at the position of interest. Indeed, the cross-section image along the yellow dotted line shows that the number of layers is dependent on the position. During the conservation treatments, maps of the mending paper pieces were recorded. By comparing Fig. 6.43d with Fig. 6.43e, the existence of the mending paper that was observed by THz pulsed TDI before conservation was proved.

The area of a previous catch on the screen is shown in Fig. 6.44a. The hole was mended using paper with gold leaf. Figure 6.44b shows the THz reflection area image and the cross-section along the yellow broken line. The gold leaf area is larger in size than the visible gold areas. The cross-section image proved that the edge of the gold leaf area was painted on. In addition to the area of the catch, the trunk of the cypress was also painted on gold leaf. The THz reflection image at a depth of 0.5 mm below the surface was extracted to determine the mending paper distribution, in the same manner as that in Fig. 6.43d. Figure 6.44c revealed that various types of mending papers were used, as expected, because this part was where a catch for the sliding door was located in the original configuration of the artwork. The mending paper map that was recorded by the conservators completely agreed with the results of THz pulsed TDI, as shown in Fig. 6.44d.

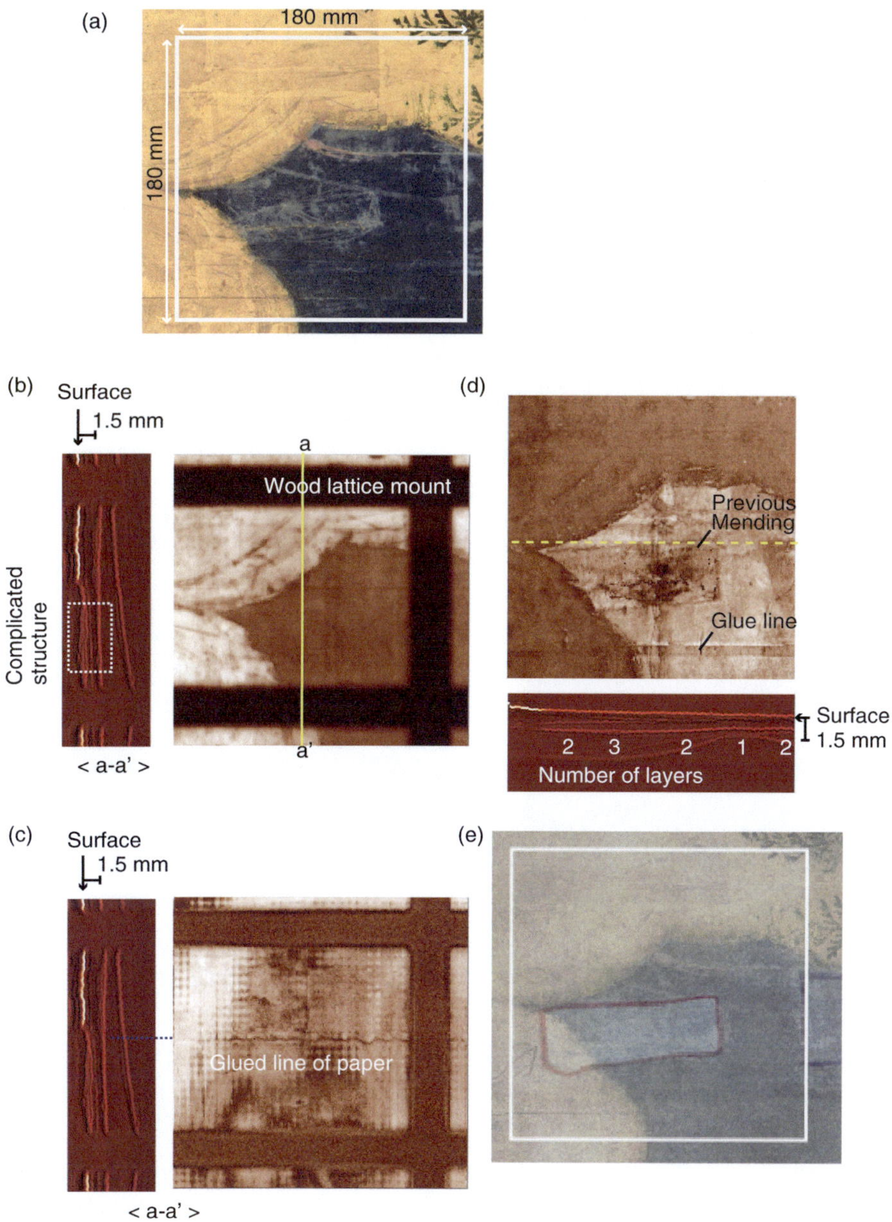

Fig. 6.43 THz pulsed TDI of part of "Cypress Tree". (**a**) Observation area, (**b**) THz reflection image from the rear side and cross-section images along the *yellow solid lines* marked *a-a′*, (**c**) THz reflection image from the rear side, (**c**) 1 mm below the surface, (**d**) THz reflection image from the front, 0.5 mm below the surface, and cross-section along the *yellow dotted line*, (**e**) map of mending paper pieces recorded by conservators during previous treatment

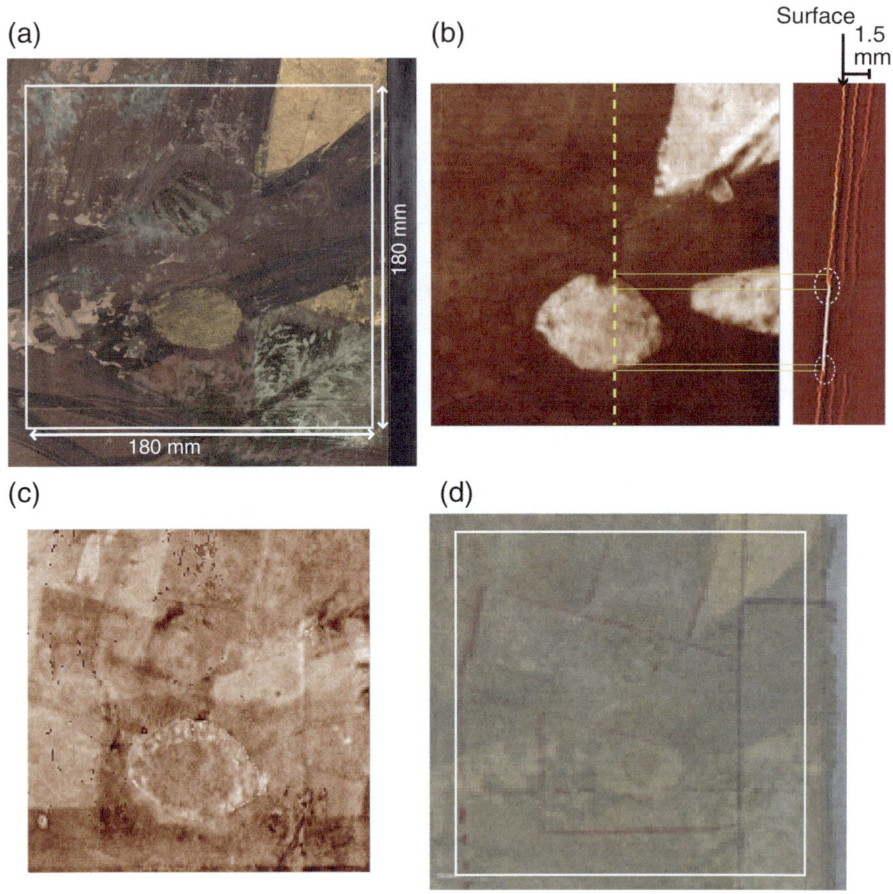

Fig. 6.44 THz pulsed TDI around a previous catch location of "Cypress Tree". (a) Observation area, (b) THz reflection image and cross-section image along the *yellow broken line*, (c) THz reflection image at a depth of around 0.5 mm below the surface, (d) map of mending papers recorded during conservation process

6.6 Oil Paintings on Canvas

Oil paintings on canvas commonly have very thin preparation and painting layers. To observe the layer structure nondestructively in a thickness of less than 0.2 mm, recent optical coherent tomography techniques may be the most effective method in various nondestructive testing methods. Obviously, various classic methods such as XRF have also been used for material analysis applications. Therefore, it has seemed that THz technology may not be suitable for use on canvas paintings.

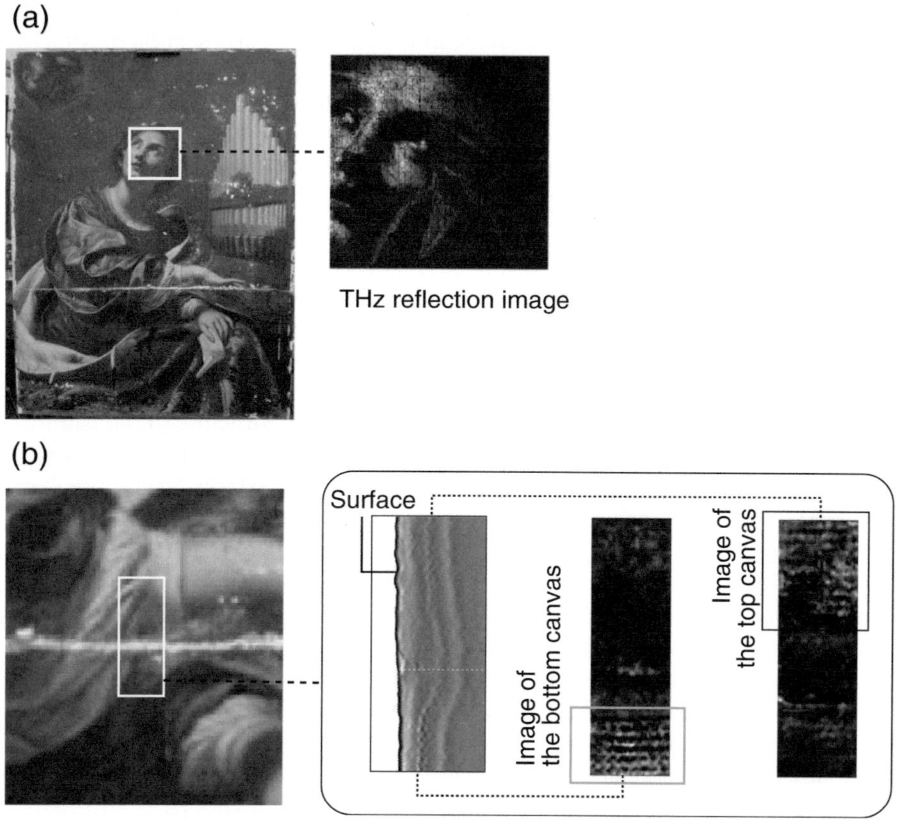

Fig. 6.45 THz pulsed TDI of an oil painting on canvas, "Santa Cecilia". (**a**) Area image around the face, (**b**) cross-section and area image at the connecting part between the two canvas sheets

However, recent works using spectroscopic imaging have produced successful results [36, 37], and there are some specific cases where THz pulsed TDI can be useful for practical conservation planning and examinations.

The first example is the THz pulsed TDI examination of the canvas itself. An oil on canvas painting, "Santa Cecilia (1626)" by Simon Vouet, was observed by THz pulsed TDI during the conservation process.

Some invisible lines were recognised in the facial area, as shown in Fig. 6.45a. These lines suggest the presence of under-painting. The pigments that were used in the artwork were not analysed, but the high reflection part of the invisible area was deduced to be painted with lead white because of its high reflectivity in the THz frequency range.

The advantage of THz pulsed TDI in this case is in the examination of the canvas. The support for this painting is made of two canvas sheets that were sewn together to make a large canvas. The sewn part had become a defect in the form of a convex edge on the surface. The conservator successfully reformed the convex part towards the backside, resulting in a concave surface, by using a suction table. As shown in Fig. 6.45b, the result is clearly visible in the THz cross-section image. The area information of the surfaces of the two canvases is also shown. The two cloths have almost the same mesh size, but the depth from the surface is slightly different. Similar to the examples shown in Figs. 6.43 and 6.44, THz pulsed TDI was able to reveal the conservation treatment that had been performed.

Another good example is the case of paintings that have been repainted, either by the artist or by conservators. Figure 6.46 shows the THz pulsed TDI of an oil painting on canvas, "Homme au Chapeau" (1915), by Pablo Picasso, which is part of the collection of the Tokyo Station Gallery.

The original canvas of the artwork was extremely thin, and it was remounted on a new thicker canvas with thick resin-type glue, in a previous conservation process. Additionally, some areas can be seen to have more than two paint layers, even by the naked eye. THz pulsed TDI was thus used to examine the structure of the painting. Figure 6.46a, b show the measurement setup and the observation area, respectively. The visible image was obtained from a catalogue from an exhibition in Tokyo, and this means that the fringe area under the artwork's frame was not shown. The THz reflection image obtained by power integration of the entire signal (Fig. 6.46c) includes information from all the layers. Some differences in the THz reflection level exist in the red area, although it appears to be uniform in the visible image.

The layer structure of this artwork is dependent on the area, as shown in Fig. 6.47. There are three different layer structures in the area indicated in Fig. 6.47a, and the cross section image of this area is shown in Fig. 6.47b. The interface between the original canvas and the new canvas is clearly observed. Area A, a pale rectangular area, has one layer with a possible separation at the edge. Here, the resolution of THz pulsed TDI is not high enough to distinguish between the paint and the original canvas. If a new paint was added to the paint before being properly cured, the layer should become a single solid layer over time, which would result in it being recognised as a single layer by THz pulsed TDI. The red areas indicated by B and C, in contrast, have complex internal structures. The original canvas and the paint layer appear as one or two layers with possible separations, indicated by C and C′, while the area indicated by B has one additional layer on top. This top layer appears in the relatively dark area shown in Fig. 6.46c. The area image at the depth of the original canvas was extracted and is shown in Fig. 6.47c. The brush strokes are clearer than those in Fig. 6.46c, and the image is not as uniform as it appears in its present condition.

However, while scientific examinations can reveal the layer structure and the materials used, the data cannot tell us who added the paint. It is well known that Picasso often repainted his works, and used various types of art materials, including industrial paint intended for carpenters [38]. Therefore, to resolve these questions, and to proceed with conservation work in the twenty first century, more historical background information is required. The THz reflection image is similar to one photograph of the work among three shown in two catalogues and a book [39–41]. If all layers were painted by the artist himself, then the earlier layers cannot have been removed. By extracting the images layer by layer, THz TDS may be able to display all artworks created by the artist.

(a)

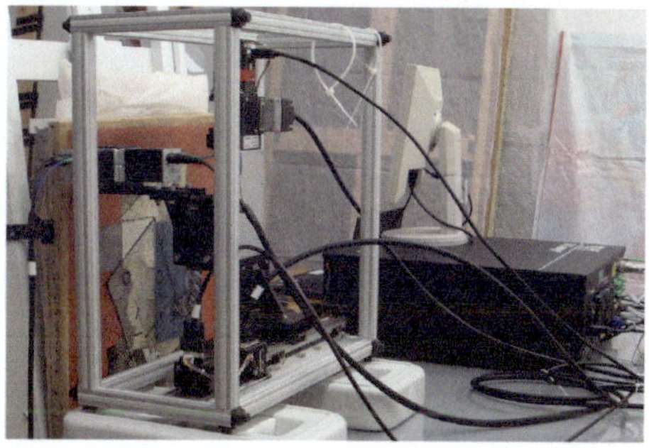

(b)

(c)

Fig. 6.46 THz pulsed TDI of an oil on canvas painting, "Homme au Chapeau". (**a**) Measurement setup, (**b**) observation area, (**c**) THz reflection image

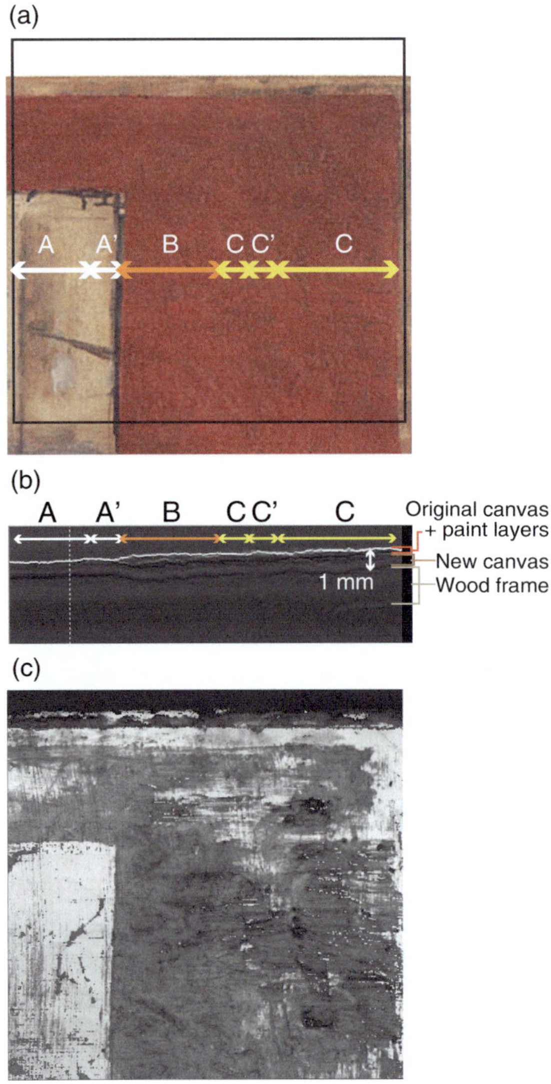

Fig. 6.47 Cross-section analysis of "Homme au Chapeau". (**a**) Visible image with indications corresponding to the layer structure, (**b**) cross-section image along the *arrow lines* shown in part (**a**), (**c**) THz reflection image from the depth of the original canvas layer

References

1. D.T. Mininberg, Neurosurgery **49**, 192 (2001)
2. H. Hoffman, W.E. Torres, R.D. Ernst, Paleoradiology: advanced CT in the evaluation of nine Egyptian mummies. Radiographics **22**, 377–385 (2002)
3. F. Cesarani, M.C. Martina, A. Ferraris, R. Grilletto, R. Boano, E.F. Marochetti, A.M. Donadoni, G. Gandini, Whole-body three-dimensional multidetector CT of 13 Egyptian human mummies. Am. J. Roentgenol. **180**, 597–606 (2003)

4. K. Fukunaga, E. Cortes, A. Cosentino, I. Stünkel, M. Leona, I.N. Duling III, D.T. Miniberg, Investigating the use of terahertz pulsed time domain reflection imaging for the study of fabric layers of an Egyptian mummy. J. Eur. Opt. Soc. Rapid Commun. **6**, 11040 (2011)
5. M.R. Stringer, D.N. Lund, A.P. Foulds, A. Uddin, E. Berry, R.E. Miles, A.G. Davies, The analysis of human cortical bone by terahertz time-domain spectroscopy. Phys. Med. Biol. **50**, 3211–3219 (2005)
6. D. Rosalie (ed.), *Egyptian Mummies and Modern Science* (Cambridge University Press, Cambridge, 2008)
7. L. Ohrström, A. Bitzer, M. Walther, F.J. Rühli, Technical note: terahertz imaging of ancient mummies and bone. Am. J. Phys. Anthropol. **142**, 497–500 (2010)
8. M. Bessou, B. Chassagne, J.-P. Caumes, C. Pradère, P. Maire, M. Tondusson, E. Abraham, Three-dimensional terahertz computed tomography of human bones. Appl. Opt. **51**(28), 6738–6744 (2012)
9. J.B. Jackson, J. Labaune, R. Bailleul-Lesuer, L. D'Alessandro, A. Whyte, J.W. Bowen, M. Menu, G. Mourou, Terahertz pulse imaging in archaeology. Front. Optoelectron. **8**(1), 81–92 (2015)
10. L. Öhrström, B.M. Fischer, A. Bitzer, J. Wallauer, M. Walther, F. Rühli, Terahertz imaging modalities of ancient Egyptian mummified objects and of a naturally mummified rat. Anat Rec. Spec. Issue: Anat. Mummy **298**(6), 1135–1143 (2015)
11. J.B. Jackson, M. Mourou, J.F. Whitaker, I.N.D. III, S.L. Williamson, M. Menu, G.A. Mourou, Terahertz imaging for non-destructive evaluation of mural paintings. Opt. Commun. **281**, 527–532 (2008)
12. G.C. Walker, J.W. Bowen, J-B. Jackson, J. Labaune, G. Mourou, M. Menu, W. Matthews, I. Hodder, "Sub-surface Terahertz Imaging through Uneven Surfaces: Visualizing Neolithic Wall Paintings in Çatalhöyük", Proc. CLEO: Science and Innovations, No. CTu3B (2012)
13. C. Cennini, "Il Libro dell'arte (ca. 1390)", Ed. F. Frezzato, Neri Pozza, Venezia (2009)
14. M. Vitruvius Pollio, "De Architecture (29-23 BCE)", ed. L. Cherubini, Giardini, Pisa (1975)
15. K. Fukunaga, T. Meldrum, W. Zia, M. Ohno, T. Fuchida, B. Blümich, Nondestructive investigation of the internal structure of fresco paintings. Proc. Digit. Herit. **1**, 81–88 (2013)
16. Agency for Cultural Affiars, *Catalogue of Exhibition "Mural paintings of the Kitora tumulus"* (Tokyo National Museum, Tokyo, 2014)
17. S.L. Renshaw, Astronomical iconography in Takamatsuzuka and Kitora tumuli: anomalies inthe adaptation of astronomical and cosmological knowledge in early Japan. Mediterr. Archaeol. Archaeometry **14**, 197–210 (2014)
18. T. Kiyuna, K.-D. An, R. Kigawa, C. Sano, S. Miura, J. Sugiyama, Bristle-like fungal colonizers on the stone walls of the Kitora and Takamatsuzuka Tumuli are identified as Kendrickiella phycomyces. Mycroscience **53**, 446–459 (2012)
19. R. Kigawa, C. Sano, M. Nishijima, N. Tazato, T. Kiyuna, N. Hayakawa, W. Kawanobe, S. Utagawa, T. Tateishi, J. Sugiyama, Investigation of acetic acid bacteria isolated form the Kitora tumulus in Japan and their involvement in the deterioration of the plaster of the mural paintings. Stud. Conserv. **58**, 30–40 (2013)
20. Y. Kohdzuma, Analysis of inorganic pigments with Laser Raman spectrsocpy, Annual Report of Nara National Research Institute for Cultural Properties, 20–21 (2002)
21. K. Fukunaga, I. Hosako, Y. Kohdzuma, T. Koezuka, M.-J. Kim, T. Ikari, X. Du, Terahertz analysis of an East Asian historical mural painting. J. Eur. Opt. Soc. Rapid Publ. **5**(10024), (2010)
22. M. Koch, S. Hunsche, P. Schumacher, M.C. Nuss, J. Feldmann, J. Fromm, THz pulsed TDI: a new method for density mapping of wood. Wood Sci. Technol. **32**, 421–427 (1998)
23. J.B. Jackson, M. Mourou, J. Labaune, J.F. Whitaker, I.N. Duling, S.L. Williamson, C. Lavier, M. Menu, and G. Mourou, Terahertz pulse imaging for tree-ring analysis: a preliminary study for dendrochronology applications. Meas. Sci. Technol. **20**, 075502 (2009)
24. K. Fukunaga, Non-destructive THz pulsed TDI of Giotto masterpiece. News Conserv. **10**, 2 (2009)

25. R.M. Groves, B. Pradarutti, E. Kouloumpi, W. Osten, G. Notni, 2D and 3D non-destructive evaluation of a wooden panel painting using shearography and terahertz imaging. NDT E Int. **42**(6), 543–549 (2009)
26. A. Tartuferi (ed.), *Giotto, Il Restauro del Polittico di Badia* (Mandoragora, Firenze, 2012)
27. M. Picollo, K. Fukunaga, J. Labaune, Obtaining noninvasive stratigraphic details of panel paintings using terahertz time domain spectroscopy imaging system. J. Cult. Herit. **16**, 73–80 (2014)
28. K. Fukunaga, M. Picollo, Terahertz time-domain imaging for cultural heritage. Insight **56**, 142–146 (2014)
29. C.L.K. Dandolo, P.U. Jepsen, M.C. Christiansen, Characterization of European lacquers by Terahertz (THz) reflectometric imaging, IEEE Proceedings of 1st digital heritage conference, 2003, pp. 89–94, 2013
30. A.S. Skryl, J.B. Jackson, M.I. Bakunov, M. Menu, G.A. Mourou, Terahertz time-domain imaging of hidden defects in wooden artworks: application to a Russian icon painting. Appl. Opt. **53**, 1033–1038 (2014)
31. K. Toishi, Exfoliation of chalk-raised paintings. Sci. Conserv. **12**, 71076 (1974). in Japanese
32. N. Kamba, Y. Tsuchiya, A. Okimoto, K. Fukunaga, Internal structure observation of a Japanese panel painted screen by Terathertz imaging technique. Proceeding 10th international conference on non-destructive investigations and microanalysis for the diagnostics and conservation of cultural and environmental heritage, NDT-41 (2011)
33. M. Leona, J. Perry, P. Londero, K. Fukunaga, G. Bailey, C. Hale, "Designing nature: Ogata Korin's technical choice in Irises at Yatsuhashi", Chapter 16 of "Science and Art, The Painted Surface" (The Royal Society of Chemistry, 2014), pp. 336–353,
34. Nezu Museum Ed., "National Treasure Irises", Nezu Museum (2005)
35. Tokyo National Museum Ed, Special issue on Conservation of National Treasure "Cypress tree", MUSEUM, No. 654, (2015). in Japanese
36. C. Seco, V. Lopez, G. Arauz, A. Redo, J. Palacios, J. Tejada, Goya's artwork imaging with terahertz waves. Opt. Express **21**, 17800–17805 (2013)
37. C.L. Koch-Dandolo, T. Filtenborg, K. Fukunaga, J. Skou-Hansen, P.U. Jepsen, Reflection terahertz time-domain imaging for analysis of an 18th century neoclassical easel painting. Appl. Opt. **54**(16), 5123–5129 (2015)
38. F. Cassadio, V. Rose, High-resolution fluorescence mapping of impurities in historical zinc oxide pigments: hard X-ray nanoprobe applications to the paints of Pablo Picasso. Appl. Phys. A **111**, 1–8 (2013)
39. C. Zervos, *Pablo Picasso par Christian Zervos*, vol. 29 (Editions Cahiers d'Art, Paris, 1975), p.68
40. P. Daix, J. Rosselet, "Le Cubisme de Picasso", (Editions Ides et Calendes, Neuchâtel, 1979), p. 352
41. J. Richardson, *A Life of Picasso* (Jonathan Cape, London, 1996), p. 350

MIX
Papier aus verantwortungsvollen Quellen
Paper from responsible sources
FSC® C105338

If you have any concerns about our products,
you can contact us on
ProductSafety@springernature.com

In case Publisher is established outside the EU,
the EU authorized representative is:
Springer Nature Customer Service Center GmbH
Europaplatz 3, 69115 Heidelberg, Germany

Printed by Libri Plureos GmbH
in Hamburg, Germany